Evaluating the Effectiveness of Interventions to Prevent and Address Sexual Harassment

PROCEEDINGS OF A WORKSHOP

Arielle L. Baker, Jeena M. Thomas,
and Jennifer E. Saunders, *Rapporteurs*

Committee on Developing Evaluation Metrics
for Sexual Harassment Prevention Efforts

Committee on Women in Science, Engineering, and Medicine

Policy and Global Affairs

The National Academies of
SCIENCES • ENGINEERING • MEDICINE

THE NATIONAL ACADEMIES PRESS
Washington, DC
www.nap.edu

THE NATIONAL ACADEMIES PRESS 500 Fifth Street, NW Washington, DC 20001

This activity was supported by the National Institutes of Health. Any opinions, findings, conclusions, or recommendations expressed in this publication do not necessarily reflect the views of any organization or agency that provided support for the project.

International Standard Book Number-13: 978-0-309-08769-8
International Standard Book Number-10: 0-309-08769-4
Digital Object Identifier: https://doi.org/10.17226/26279

Additional copies of this publication are available from the National Academies Press, 500 Fifth Street, NW, Keck 360, Washington, DC 20001; (800) 624-6242 or (202) 334-3313; http://www.nap.edu.

Copyright 2021 by the National Academy of Sciences. All rights reserved.

Printed in the United States of America

Suggested citation: National Academies of Sciences, Engineering, and Medicine. 2021. *Evaluating the Effectiveness of Interventions to Prevent and Address Sexual Harassment: Proceedings of a Workshop*. Washington, DC: The National Academies Press. https://doi.org/10.17226/26279.

The National Academies of
SCIENCES • ENGINEERING • MEDICINE

The **National Academy of Sciences** was established in 1863 by an Act of Congress, signed by President Lincoln, as a private, nongovernmental institution to advise the nation on issues related to science and technology. Members are elected by their peers for outstanding contributions to research. Dr. Marcia McNutt is president.

The **National Academy of Engineering** was established in 1964 under the charter of the National Academy of Sciences to bring the practices of engineering to advising the nation. Members are elected by their peers for extraordinary contributions to engineering. Dr. C. D. Mote, Jr., is president.

The **National Academy of Medicine** (formerly the Institute of Medicine) was established in 1970 under the charter of the National Academy of Sciences to advise the nation on medical and health issues. Members are elected by their peers for distinguished contributions to medicine and health. Dr. Victor J. Dzau is president.

The three Academies work together as the **National Academies of Sciences, Engineering, and Medicine** to provide independent, objective analysis and advice to the nation and conduct other activities to solve complex problems and inform public policy decisions. The National Academies also encourage education and research, recognize outstanding contributions to knowledge, and increase public understanding in matters of science, engineering, and medicine.

Learn more about the National Academies of Sciences, Engineering, and Medicine at **www.nationalacademies.org**.

The National Academies of
SCIENCES · ENGINEERING · MEDICINE

Consensus Study Reports published by the National Academies of Sciences, Engineering, and Medicine document the evidence-based consensus on the study's statement of task by an authoring committee of experts. Reports typically include findings, conclusions, and recommendations based on information gathered by the committee and the committee's deliberations. Each report has been subjected to a rigorous and independent peer-review process and it represents the position of the National Academies on the statement of task.

Proceedings published by the National Academies of Sciences, Engineering, and Medicine chronicle the presentations and discussions at a workshop, symposium, or other event convened by the National Academies. The statements and opinions contained in proceedings are those of the participants and are not endorsed by other participants, the planning committee, or the National Academies.

For information about other products and activities of the National Academies, please visit www.nationalacademies.org/about/whatwedo.

PLANNING COMMITTEE ON DEVELOPING EVALUATION METRICS FOR SEXUAL HARASSMENT PREVENTION EFFORTS

Vicki J. Magley (*Chair*), Professor, Department of Psychological Sciences, University of Connecticut
NiCole T. Buchanan, Professor, Department of Psychology, Michigan State University
Carol W. Greider, Professor of Molecular, Cellular, and Developmental Biology, University of California, Santa Cruz
Melissa L. Kwon, Associate Director for Prevention, PATH to Care Center, University of California, Berkeley
Larry R. Martinez, Associate Professor and Associate Chair, Department of Psychology, Portland State University
Nicole M. Merhill, Director, Office for Gender Equity and University Title IX Coordinator, Harvard University

Project Staff

Arielle L. Baker, Program Officer, Committee on Women in Science, Engineering, and Medicine, the National Academies
Frazier Benya, Senior Program Officer, Committee on Women in Science, Engineering, and Medicine, the National Academies
Imani Braxton-Allen, Senior Program Assistant, Committee on Women in Science, Engineering, and Medicine, the National Academies (until April 2021)
Abigail Harless, Senior Program Assistant, Committee on Women in Science, Engineering, and Medicine, the National Academies (beginning June 2021)
Jeena M. Thomas, Program Officer, Committee on Women in Science, Engineering, and Medicine, the National Academies
Jen Saunders, Consultant Writer

Preface and Acknowledgments

Higher education institutions have been carrying out efforts to prevent and address sexual harassment for decades. Yet there is still little known about whether those prevention efforts, or the efforts to evaluate preventative measures, are effective in achieving their goals. Until now, a common space to share perspectives about and make progress on such evaluation efforts—on what institutions are doing, on what challenges practitioners are facing, on what research needs to be done, on whether sexual harassment is being effectively prevented—has not been made available.

This Proceedings of a Workshop was prepared by the workshop rapporteurs as a factual summary of what was presented and discussed at the workshop. The role of the planning committee was limited to planning and convening the workshop. The statements made are those of the rapporteurs and do not necessarily represent positions of the workshop participants as a whole, the planning committee, or the National Academies of Sciences, Engineering, and Medicine. I extend sincere thanks to the members of the planning committee for their contributions in scoping, developing, and carrying out this project. I also thank staff members Layne Scherer and Maria Lund Dahlberg, who supported the execution of this workshop. The workshop was made possible thanks to support from the National Institutes of Health.

This Proceedings of a Workshop was reviewed in draft form by individuals chosen for their diverse perspectives and technical expertise. The purpose of this independent review is to provide candid and critical comments that will assist the National Academies of Sciences, Engineering, and Medicine in making each published proceedings as sound as possible and to ensure that it meets the institutional standards for quality, objectivity, evidence, and responsiveness to the charge. The review comments and draft manuscript remain confidential to protect the integrity of the process. We wish to thank the following individuals for their review of this proceedings: Connie Citro, The National Academies of Sciences, Engineering, and Medicine; Jennifer Jacobsen, Macalester College; Dana Kabat-Farr, Dalhousie University; Kurt Kraiger, University of Memphis; and Vicki Magley, University of Connecticut. Although the reviewers listed above have provided many constructive comments and suggestions, they were not asked to endorse the content of the proceedings, nor did they see the final draft before its release. The review of this proceedings was overseen by Timothy Johnson, University of Michigan. He was responsible for making certain that an independent examination of this proceedings was carried out in accordance with standards of the National Academies and that all review comments were carefully considered. Responsibility for the final content rests entirely with the rapporteurs and the National Academies.

> Arielle L. Baker, Program Officer
> Committee on Women in Science, Engineering, and Medicine
> National Academies of Sciences, Engineering, and Medicine

Contents

1 INTRODUCTION — 1

2 SETTING THE STAGE: EVALUATING EFFORTS TO PREVENT AND ADDRESS SEXUAL HARASSMENT — 5
Steps in Program Evaluation, 6
Existing Measures and Metrics for Evaluating Change in Organizational Climate, 8

3 USING PREVENTION SCIENCE AND IMPLEMENTATION SCIENCE TO BETTER EVALUATE SEXUAL HARASSMENT PREVENTION EFFORTS — 13
Applying Prevention Science, 13
Applying Implementation Science, 19

4 CHALLENGES AND LIMITATIONS THAT ARISE WHEN EVALUATING SEXUAL HARASSMENT PREVENTION EFFORTS — 25

5. PARTICIPANT REFLECTIONS ON WORKSHOP PRESENTATIONS AND DISCUSSIONS — 29
Taking a Systemic and Outcome-Focused Approach, 29
Considering Those Most Vulnerable to Harassment, 32

Leadership as a Linchpin: Building Support
      and Communicating Value, 34
   Expanding the Evaluation Toolkit, 35
   Building Community, Collaboration, and Trust, 37

6 EVALUATION IN ACTION: EXAMPLES
   AND RESOURCES   39
   Examples of Efforts to Evaluate Sexual Harassment
      Interventions, 39
   Applying Implementation Science to Understand
      Contextual Factors that Facilitate or Impede
      Successful Efforts, 44

7 REFLECTIONS BY EDEN KING ON THEMES
   AND NEXT STEPS   47
   The Unique Challenge, 47
   Thematic Synthesis, 48
   Reflections for Moving Forward, 50
   Next Steps, 51

REFERENCES   53

APPENDIXES

A WORKSHOP AGENDA   57
B BIOGRAPHICAL SKETCHES OF PLANNING
   COMMITTEE MEMBERS, SPEAKERS,
   AND MODERATORS   67
C WORKSHOP PARTICIPANTS   79
D WORKSHOP CASE STUDIES   85
E WORKSHEETS FOR GETTING STARTED
   WITH IMPLEMENTATION SCIENCE   111

1

Introduction

Recent years have brought to light a long-standing and pervasive issue in academia—sexual harassment. A report of the National Academies of Sciences, Engineering, and Medicine (NASEM, 2018) (hereafter, National Academies) found disturbingly high rates of sexual harassment in academia: more than 50 percent of female faculty and staff and between 20 and 50 percent of female students in higher education institutions reported experiencing sexually harassing behavior. The devastating impact of sexual harassment, both to those affected and the institutions, has highlighted the need for system-wide changes to the culture and climate in higher education to prevent and effectively respond to sexual harassment. As the 2018 National Academies report notes, the "cumulative effect of sexual harassment is significant damage to research integrity and a costly loss of talent in academic sciences, engineering, and medicine."

Rising awareness of and increased attention to sexual harassment has resulted in momentum to implement sexual harassment prevention efforts in higher education institutions. For example, the National Academies' Action Collaborative on Preventing Sexual Harassment in Higher Education (hereafter, Action Collaborative) is an activity of more than 60 colleges, universities, and other research and training institutions that are identifying, researching, developing, and implementing efforts that move beyond basic legal compliance to evidence-based policies and practices for addressing and preventing all forms of sexual harassment and promoting a campus climate of civility and respect. Importantly, this effort expands beyond the 2018 report to push for targeted, collective action across all disciplines and among all people in higher education. The Action Collaborative produced a Year 1 Annual Report (NASEM, 2020a) highlighting overall progress and summarizing members' actions during the first year. As that report notes, work on preventing sexual harassment is an area that has recently garnered a lot of attention, especially around education and programs that go beyond the standard anti-sexual harassment trainings often used to comply with legal requirements. The annual report states that "although increased attention in this space is encouraging, there is a significant need for evaluation of the effectiveness of such prevention efforts, but also over a longer time span."

On April 20–21, 2021, the National Academies hosted a workshop aimed at tackling this issue: see Box 1-1 for the statement of task. Due to public health restrictions on in-person gatherings related to the COVID-19 pandemic, this workshop was held virtually. The workshop included discussion around approaches and strategies for evaluating and measuring the effectiveness of sexual harassment interventions being implemented at higher education institutions and research and training sites to assist institutions in transforming promising ideas into evidence-based best practices. Workshop participants also addressed methods, metrics, and measures that could be used to evaluate sexual harassment prevention efforts.

BOX 1-1
Workshop Statement of Task

The National Academies of Sciences, Engineering, and Medicine will conduct a 2-day Workshop on Developing Evaluation Metrics for Sexual Harassment Prevention Efforts. This workshop will discuss approaches and strategies for evaluating and measuring the effectiveness of sexual harassment interventions being implemented at higher education institutions and research and training sites, in order to assist institutions in transforming promising ideas into evidence-based best practices. The workshop discussions will focus on parameters that go beyond decreases in the prevalence of sexual harassment and to methods, metrics, and interim measures that can demonstrate a change in the organizational climate and culture and/or a change in behavior among community members.

This workshop proceedings largely follows the organization of the workshop. Following this introduction, Chapter 2 sets the stage, summarizing what is known about the current state of evaluation efforts to assess sexual harassment prevention activities in higher education. Chapter 3 provides an overview of prevention science and implementation science and their application to evaluating sexual harassment prevention. Chapter 4 includes a summary of challenges, barriers, and limitations related to evaluating sexual harassment prevention efforts. Chapter 5 offers participants' reflections from the workshop presentations and discussion, and Chapter 6 discusses examples of on-the-ground intervention evaluation work and describes relevant resources related to implementation science. Finally, Chapter 7 provides expert reflections on the workshop by industrial-organizational psychologist Eden King of Rice University, which was commissioned by the planning committee. Box 1-2 presents the definitions of key terms used in this proceedings.

Appendixes A, B, and C are the workshop agenda, biographical sketches of committee members and presenters, and a list of workshop participants. Appendix D is a set of case studies that were provided prior to the workshop, and Appendix E is a worksheet of resources for applying implementation science models to the evaluation of sexual harassment prevention efforts.

Throughout the workshop, participants considered the presentations and discussions within the context of six case studies in Appendix D: examples of actual programs, policies, or practices currently being carried out in higher education institutions around the country. As a part of their commitment to the Action Collaborative, member institutions annually submit these "Descriptions of Work," which are then made publicly available. Each participant was assigned to one of the six case studies and asked to review the description in advance of the workshop. The purpose of the case studies was to help participants consider how actual prevention efforts might be evaluated for effectiveness.

> **BOX 1-2**
> **Definition of Key Terms**
>
> **Sexual harassment** is a form of discrimination that includes gender harassment (sexist hostility and crude behavior), unwanted sexual attention (unwelcome verbal or physical sexual advances), and sexual coercion (when favorable professional or educational treatment is conditioned on sexual activity). Gender harassment is by far the most common form of sexual harassment, and when severe or frequent, it can result in the same level of negative outcomes as one instance of sexual coercion (NASEM, 2018).
>
> **Man** and **woman** refer to any person who identifies as such, including but not limited to cisgender, transgender, and non-binary individuals. Although gender is a spectrum that expands beyond these constructs (NASEM, 2020b), this document does contain some binary language to remain consistent with the remarks of workshop participants and the methodologies of the work that was presented.
>
> **Incivility** refers to "low-intensity deviant behavior with ambiguous intent to harm the target, in violation of workplace norms for mutual respect. Uncivil behaviors are characteristically rude and discourteous, displaying a lack of regard for others" (Andersson and Pearson, 1999). This term carries a history of harm that includes colonialism, erasure, displacement, and racism (Cortina et al., 2021); these shortcomings are acknowledged, but the use of the term has been retained when relaying participant remarks and presentations.
>
> **Equity** refers to an environment where "avoidable or remediable differences among groups of people, whether those groups are defined socially, economically, demographically or geographically" (World Health Organization, 2021) are removed, for instance by influencing policies or deciding on outcomes that account for such differences.

2

Setting the Stage: Evaluating Efforts to Prevent and Address Sexual Harassment

The workshop began with a discussion about what is currently known about how higher education institutions are approaching sexual harassment prevention and evaluation, including the reasons that an institution would or should evaluate its prevention efforts.

Elissa Perry, Columbia University, provided an overview of the evaluation of sexual harassment prevention efforts in higher education, based on a paper that was commissioned from her by the workshop planning committee.[1] Her paper describes sexual harassment prevalence and the importance of prevention and evaluation work, addresses how higher education institutions are approaching sexual harassment prevention work, identifies barriers to the evaluation of sexual harassment prevention efforts, and offers evidence-based suggestions for the evaluation process. Importantly, this paper highlights that strategies for addressing sexual harassment expand beyond individual- and group-level efforts, as well as traditional approaches to prevention, such as training and education. For example, institutional factors, including leadership, organizational structure, practices, and systems (not directly related to victim support or complaint procedures), may also play a role in preventing sexual harassment.

Research supports the idea that prevention programs are most effective when they are continuously evaluated, stated Perry. Evaluations yield information that can be used to justify allocations of resources and time, assess

---

[1] Available at: https://www.nap.edu/catalog/26279.

whether the intervention is having the intended effects, and determine whether it should be revised or discontinued. Training transfer, or applying knowledge acquired during training to a targeted job or role, is also more likely when the training is frequently evaluated and its effects on behavior and organizational performance are assessed.

Perry noted that higher education institutions and other organizations that have not engaged in evaluation initiatives often cite a lack of resources. Other reasons for not conducting evaluation have included the assumption that the intervention is effective, a lack of agreement on what should be evaluated, and limited knowledge about how to conduct evaluations. Importantly, there is often a disconnect between what is measured and the outcomes the intervention is intended to affect.

STEPS IN PROGRAM EVALUATION

Perry also provided an overview of a generalized program evaluation process, the first step being to conduct a needs assessment, involving diagnosing who needs to be trained, for what, and when. A needs assessment can provide information and understanding of the specific needs, what needs the intervention is designed to address (e.g., greater awareness, changed behavior, improved climate), and clarity around the outcomes the intervention should be designed to affect. It should be tied directly to the evaluation plan, stated Perry.

The second step is to select appropriate program measures. Formative evaluations focus on improving the quality of the program, including its delivery and design, while summative evaluations, which are typically conducted later in the life cycle of a program, can be used to provide evidence of the effects of the program. Perry noted that the selection of appropriate program outcome measures should be based on needs assessments. One should also consider how quickly evaluation information is needed and how long it will take for the program to affect the outcomes of interest, stated Perry.

The third step is to select an appropriate evaluation design. While randomized control trials are considered the "gold standard" of evaluation, Perry noted, they can be difficult, if not impossible, to implement in the real world. Instead, quasi-experimental designs (e.g., pre-test/post-test, time-series designs) can be used. In noting where to start, she added that it is important to use a confirmatory evaluation approach. As described in her paper, a confirmatory evaluation approach "is based on a clear theory of how the program is expected to impact the outcomes of interest and looks for and interprets patterns of relationships based on the theory," and it may help strengthen support for a causal relationship between program participation and targeted outcomes.

Evaluating complex social programs, such as sexual harassment interventions, requires a different approach than evaluating less complex programs because such programs include multiple intervention strategies for multiple stakeholders, are delivered by human agents, and are implemented in complex institutions, Perry stated. To address this, one approach may be to develop a logic model or program theory that identifies program-related constructs and maps relationships between these and program outcomes. The logic model should include context, inputs, processes, and outcomes (see example in Figure 2-1). Another approach is to employ a multi-phased, mixed-methods approach.

Many institutions are collecting an extraordinary amount of data related to sexual harassment, including the impact of prevention interventions. While climate surveys are used frequently to assess the prevalence of sexual harassment, students' knowledge and attitudes related to policies and resources, and student satisfaction after using campus resources, there are many other tools that can be used to support

> We can also design surveys to include collection of data that can be incorporated into the intervention itself, so the survey is also a driver of change.
>
> —Alan Berkowitz
> (independent consultant)

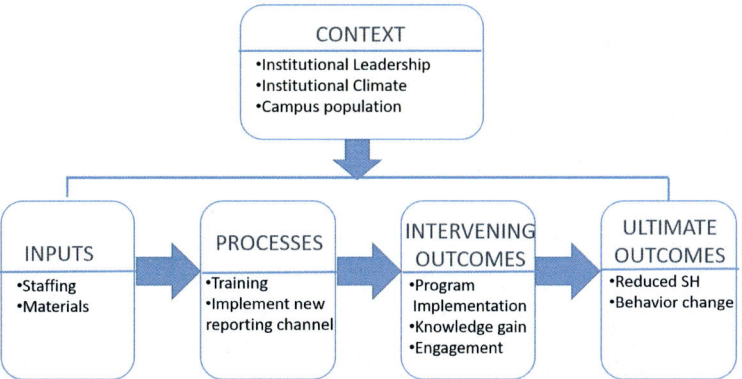

FIGURE 2-1 Sample logic model/program theory. "SH" denotes "sexual harassment".
SOURCE: Elissa Perry (2021). Workshop presentation.

program evaluation. Perry stated that higher education institutions should consider whether data collection and evaluation are occurring in a systemic fashion, for example, by conducting needs assessments. Another consideration, she noted, is whether program interventions and evaluations are being driven by the logic model or program theories that were designed to outline the key program-related activities and outcomes.

EXISTING MEASURES AND METRICS FOR EVALUATING CHANGE IN ORGANIZATIONAL CLIMATE

Several presenters noted that having measures and metrics that align with program goals is essential to the evaluation of sexual harassment prevention programs. Several speakers discussed examples of existing measures that might be used in sexual harassment prevention evaluation.

Emily Huang, Oregon Health and Science University, presented an overview of safety climate as an indicator of workplace safety (see Box 2-1). She began by defining safety culture as shared core values and beliefs that interact with an organization's structures and control systems to produce behavioral norms. Safety climate is a measurable aspect of safety culture, or the employees' perceptions of an institution's safety policies, procedures, and practices. It represents the overall importance and "true" priority of safety at work. Studies show that when a company cares about employees they have a higher level of job satisfaction. The key dimension is managerial commitment to safety, requiring internal consistency among policies, procedures, and practices. Safety climate can serve as a robust predictor of future injury, Huang noted. This approach allows for a baseline from which to measure impact. Reducing sexual harassment in the workplace contributes directly to workplace safety for employees; as one workshop participant noted, "I see diversity and inclusion as a form of safety."

Huang provided an example of a scale that captures safety culture from the perspective of management and employees (see Figure 2-2). The scale, which can also be used to assess sexual harassment safety, measures how management may or may not be working to improve safety levels or providing information on safety issues. It also assesses the level to which a direct supervisor may be engaged in discussions with an employee about how to improve safety or whether employees are following safety rules. The scale may be applicable to efforts to evaluate sexual harassment. Huang

> **BOX 2-1**
> **Definitions of Workplace Safety, Safety Culture, and Safety Climate**
>
> **Workplace safety** refers to a work environment that ensures the safety, health, and well-being of employees.
>
> **Safety culture** is defined as shared core values and beliefs that interact with an organization's structures and control systems to produce behavioral norms.
>
> **Safety climate** is a measurable aspect of safety culture, or the employees' perceptions of the safety policies, procedures, and practices. It represents the overall importance and "true" priority of safety at work.

Top management at this company…	Strongly Disagree				Strongly Agree
1. tries to continually improve safety levels in each department.	1	2	3	4	5
2. requires each manager to help improve safety in his or her department.	1	2	3	4	5
3. uses any available information to improve existing safety rules.	1	2	3	4	5
4. provides workers with a lot of information on safety issues.	1	2	3	4	5
My direct supervisor…	Strongly Disagree				Strongly Agree
1. discusses how to improve safety with us.	1	2	3	4	5
2. uses explanations (not just compliance) to get us to act safely.	1	2	3	4	5
3. reminds workers who need reminders to work safely.	1	2	3	4	5
4. makes sure we follow all the safety rules (not just the most important ones).	1	2	3	4	5

FIGURE 2-2 An item-response theory approach to safety climate measurement. SOURCE: Emily Huang (2021). Workshop presentation. Adapted from Huang et al. 2017.

added that people want to see leaders leading by example to ensure inclusive workplaces.

Another example of a measure that could be applied to evaluation of sexual harassment prevention is the masculinity contest culture (MCC) scale (see Figure 2-3). Introducing this measure, Jennifer Berdahl, University of British Columbia, said that research indicates that men are most likely to harass when masculinity in the culture is highly valued and masculinity is threatened. Harassment targets are disproportionately those who threaten the perpetrator's masculinity. Work environments that motivate men to value, prove, and defend their masculinity also motivate them to harass others based on sex.

To develop this measure, Berdahl explained, experts from a range of fields generated over 130 potential items to capture masculinity culture and refined items with online samples. A four-factor scale was developed to address identified ways to prove manhood at work: dog-eat-dog (ruthless competition); show no weakness (emotions); strength and stamina (physicality); and put work first. Respondents were asked to rank their work environment on a scale from 1 (not at all true of my work environment) to

Dog-Eat-Dog	Show No Weakness	Strength and Stamina	Put Work First
1. You're either "in" or you're "out," and once you're out, you're out	2. Admitting you don't know the answer looks weak	3. It's important to be in good physical shape to be respected	4. To succeed you can't let family interfere with work
5. If you don't stand up for yourself people will step on you	6. Expressing any emotion other than anger or pride is seen as weak	7. People who are physically smaller have to work harder to get respect	8. Taking days off is frowned upon
9. You can't be too trusting	10. Seeking other's advice is seen as weak	11. Physically imposing people have more influence	12. To get ahead you need to be able to work long hours
13. You've got to watch your back	14. The most respected people don't show emotions	15. Physical stamina is admired	16. Leadership expects employees to put work first
17. One person's loss is another person's gain	18. People who show doubt lose respect	19. Athletic people are especially admired	20. People with significant demands outside of work don't make it very far

FIGURE 2-3 Masculinity contest culture (MCC) scale: selected items. The blue shading of the table indicates shortened versions of the scale. The 8-item version is contained in the first two rows (#1-8), the 12-item version in the first three rows (#1-12), and the full version across all five rows (#1-20).
SOURCE: Jennifer Berdahl (2021). Workshop presentation. Adapted from Glick, Berdahl, and Alonso (2018).

5 (entirely true of my work environment). The MCC scale was found to be higher in more male-dominated organizations.

The MCC scale also predicts negative organizational and individual outcomes for both men and women. Berdahl notes that correlations from 0.5 to 0.7 indicate that there may be few women in management, toxic leadership, and identity-based harassment with low psychological safety. In environments with correlations from 0.3 to 0.5, there is a higher likelihood of individual poor personal performance, poor mental health, burnout, and job dissatisfaction, Berdahl said (see Figure 2-4; also Glick et al., 2018).

Alec Smidt, Yale University School of Medicine, discussed measures of institutional betrayal, defined as the failure to prevent harm or to respond supportively to wrongdoings when those within the institution (such as staff, students, and faculty) have a reasonable expectation of not being harmed. Smidt described studies that indicate that college women who were sexually assaulted experienced worse psychological and physical health outcomes following the trauma (see Cortina et al., 1998; Huerta et al.,

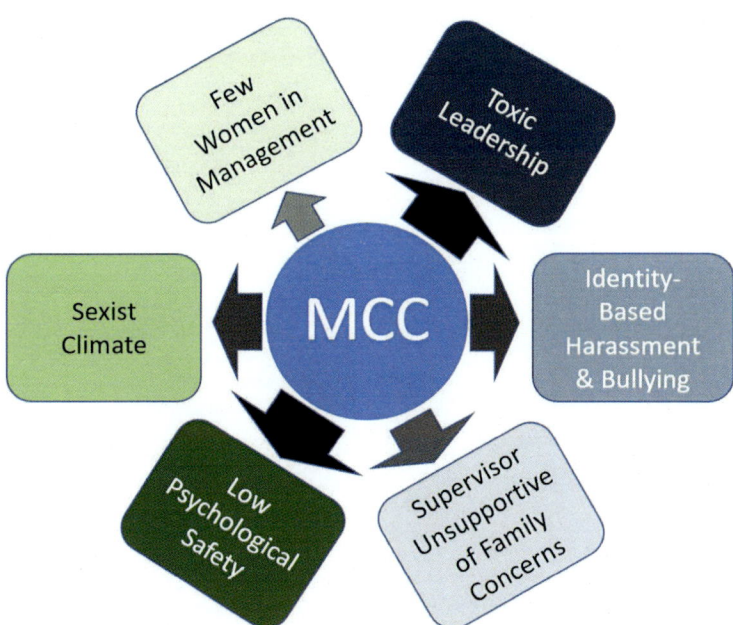

FIGURE 2-4 Organizational outcomes for both men and women in a masculinity contest culture (MCC) scale with correlations from 0.5 to 0.7.
SOURCE: Jennifer Berdahl (2021). Workshop presentation.

2006). Sexual harassment by those in powerful positions was more likely to result in institutional betrayal. By contrast, institutional courage is defined as action taken on behalf of an institution to address harassment and support those affected; as one participant noted, institutional courage implies institutional risk. Smidt discussed the institutional betrayal and support questionnaire (IBSQ), which was used in 2014 at the University of Oregon (see Figure 2-5). As described above, existing measures and predictors of organizational outcomes could be used when evaluating efforts to address sexual harassment and to understand their effectiveness and impact on organizational climate.

Following the presentations, participants discussed how these measures and others might be applicable in their work and the importance of aligning measures with program goals. For example, participants noted the importance of ongoing stakeholder engagement throughout an evaluation. Also, understanding the ability of stakeholders to implement the program was identified as an item of interest that might be measured. Regarding the MCC scale, Diana Lautenberger, Association of American Medical Colleges, noted that it would be helpful to assess the willingness of men and others to intervene, be allies, and challenge other men in their behavior. Additionally, Vicki Magley, University of Connecticut, noted that while measures of institutional betrayal and support or courage appear to be focused on those who are directly targeted by sexual harassment, perhaps even requiring reporting, these could also be applied more broadly at the institutional level.

SUPPORT	BETRAYAL
2. Apologizing for what happened to you?	10. Creating an environment in which this type of experience/s seemed common or normal?
3. Believing your report?	11. Creating an environment in which this experience seemed more likely to occur?
4. Allowing you to have a say in how your report was handled?	12. Making it difficult to report the experience/s?
6. Meeting your needs for support and accommodations	13. Responding inadequately to the experience/s, if reported?
7. Create an environment where this type of experience was safe to discuss?	14. Mishandling your case, if disciplinary action was requested?
8. Create an environment where this type of experience was recognized as a problem?	15. Covering up the experience/s?
	16. Denying your experience/s in some way?
	17. Punishing you in some way for reporting the experience/s (e.g., loss of privileges or status)?
	18. Suggesting your experience/s might affect the reputation of the institution?

FIGURE 2-5 Institutional betrayal and support questionnaire: selected items.
SOURCE: Alec Smidt (2021). Workshop presentation. Based on Smith and Freyd (2013); first reported in Rosenthal, Smidt, and Freyd (2016).

3

Using Prevention Science and Implementation Science to Better Evaluate Sexual Harassment Prevention Efforts

This chapter covers the content presented on the principles of prevention science and implementation science, including how they might be applied to the evaluation of sexual harassment prevention efforts.

APPLYING PREVENTION SCIENCE

Cindy Crusto, Yale University School of Medicine, and Lisa Hooper, University of Northern Iowa, presented key tenets of prevention science and program evaluation as a proposed framework for preventing sexual harassment in higher education. This information was based on their paper outlining how prevention science has been and is currently being applied to sexual harassment evaluation in higher education, commissioned by the workshop planning committee.[1] The paper also offers an organizing pre-

> **Prevention science** focuses on the development, implementation, and evaluation of evidence-based programs and strategies that reduce risk factors and enhance protective factors to improve the health and well-being of individuals, families, communities, and organizations.

---

[1] Available at: https://www.nap.edu/catalog/26279.

vention evaluation framework for sexual harassment for use in diverse higher education contexts.

Hooper noted that prevention science focuses on the development, implementation, and evaluation of evidence-based programs and strategies that reduce risk factors and enhance protective factors to improve the health and well-being of individuals, families, communities, and organizations. Drawing from a diverse range of disciplines, prevention science aims to understand the determinants of societal, community, and individual level problems, such as sexual harassment. Prevention science also examines risk factors, which are associated with increasing the likelihood of developing the priority problem; protective factors, which are associated with reducing the likelihood of developing the problem; and promotive factors, which are associated with optimizing health rather than protecting health.

To select and develop a prevention framework, Hooper discussed the need to consider:

- accumulated evidence, by identifying prevention programs and interventions with empirical support that meet the organizational need to prevent the problem;
- conceptual fit, through a review of available programs and interventions that appear to meet the institutional need to prevent the problem; and
- practical fit, by reviewing available programs and interventions that appear to align with the targeted population and type of institution.

Commonly used principles to guide prevention program development and process include:

- assessing available empirical support,
- understanding historical efforts,
- establishing a theory of change or logic model,
- clearly identifying the priority problem,
- identifying risk and protective factors and prevention targets, and
- developing short- and intermediate-term outcomes.

Hooper presented on a prevention science program development framework developed by the Substance Abuse and Mental Health Services Administration (see Figure 3-1). The agency describes two relevant guiding

Step 1: Assessment	Step 2: Capacity Building	Step 3: Planning	Step 4: Implementation	Step 5: Evaluation
Assess problems and related behaviors	Engage organizational stakeholders	Prioritize protective and risk factors	Deliver programs and practices	Conduct process evaluation
Prioritize problems (magnitude, trends, severity, comparison)	Develop and strengthen a prevention team	Select prevention interventions with empirical support and organizational fit	Balance fidelity with flexibility and necessary adaptations	Conduct outcome evaluation
Assess risk and protective factors	Raise organizational awareness	Develop a plan that is consistent with a logic model	Retain core components	Disseminate evaluation outcomes
			Establish implementation supports	Make improvements
Principles that Undergird the framework: Cultural Competence, and Sustainability				

FIGURE 3-1 Prevention science program development framework.
SOURCE: Cindy Crusto and Lisa Hooper (2021). Workshop presentation. Adapted from the Substance Abuse and Mental Health Services Administration (2019).

principles for the framework: (1) cultural competence, defined as "the ability of an individual or organization to understand and interact effectively with people who have different values, lifestyles, and traditions based on their distinctive heritage and social relationships," and (2) sustainability, defined as "the process of building an adaptive and effective system that achieves and maintains desired long-term results."

Hooper also presented another prevention framework for sexual harassment (see Figure 3-2). Factors that form the foundation of the framework include a focus on well-being and a culturally responsible, safe, and supportive climate and organization. Other key features include assessment, evaluation, prevention, and response practices; leadership and organizational support; and stakeholders who are trained to be culturally responsive, among other areas.

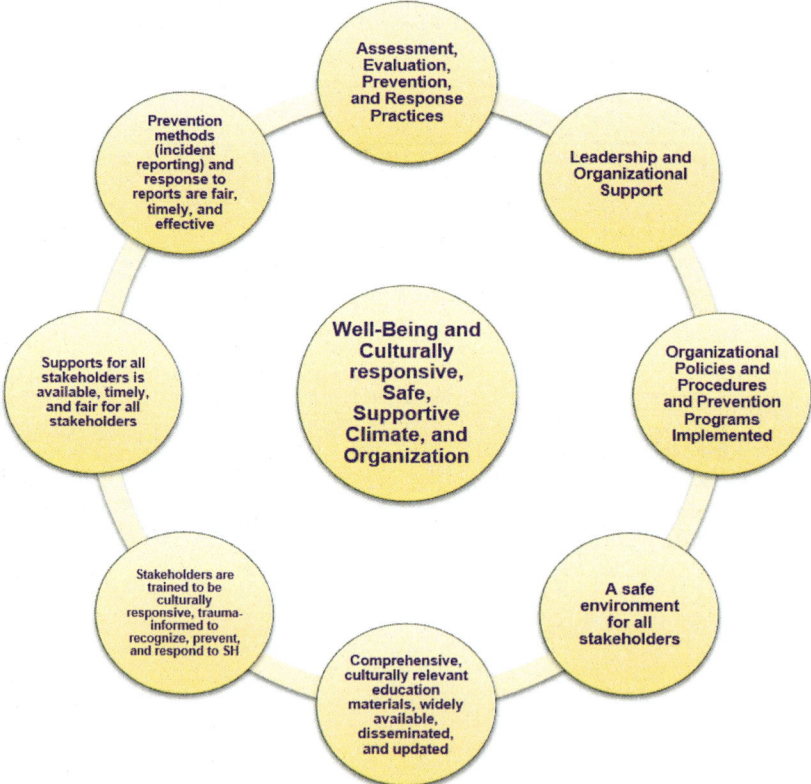

FIGURE 3-2 Prevention framework for sexual harassment.
SOURCE: Cindy Crusto and Lisa Hooper (2021). Workshop presentation.

The program evaluation process should begin at program development and continue through program implementation and beyond, stated Crusto. Developing appropriate outcomes is a critical part of the evaluation process. Outcomes can also reflect changes in learning, action, and condition. Not all outcomes occur at the same time, and some are necessary before others can happen. There are, for example, short-term, intermediate, and long-term outcomes.

To gain buy-in from participants during an evaluation, Crusto recommended including participants in the process from the beginning and asking evaluation questions that are important to them. The goal is to build an ongoing learning program or culture; the capacity to engage participants is key to achieving this goal. It is also critical to build in equity from the beginning of the program. Crusto added that the prevention lens can be helpful for understanding protective factors, including when to intervene, which can occur at the individual or systems level, and can help shape action on how to modify or adapt.

> "Matching an intervention to a community's level of readiness is absolutely essential for success. Interventions must be challenging enough to move a community forward in its level of readiness. However, efforts that are too ambitious are likely to fail because community members will not be ready or able to respond. To maximize chances for success, the Community Readiness Model offers tools to measure readiness and to develop stage-appropriate strategies."
>
> SOURCE: Plested et al., 2006

The Community Readiness Model was also discussed as a tool to assess whether a community is prepared to take action on an issue. Crusto described nine stages of community readiness ranging from no awareness (Stage 1) to high level of community ownership (Stage 9). There are also opportunities to "create" readiness by carrying out community training before conducting the intervention.

Crusto also presented the application of a logic model to an initiative at Rutgers University aimed at significantly increasing faculty and staff education and skill development.[2] The case study captured the project's goals, outputs, and outcomes. Considering inputs or outputs can support training that is trauma-informed and is faculty, staff, and student-focused. As was

---

[2] Available at: https://www.nap.edu/catalog/26279.

discussed, the case study clarifies the importance of clearly communicating expectations through university policies.

Regarding developing training programs, we know what works in prevention—a "one and done" model does not suffice in terms of resulting in long-term knowledge or behavior change, stated Crusto. At a minimum, booster sessions and follow-up assessment are needed over time.

APPLYING IMPLEMENTATION SCIENCE

Raechel Soicher and Kathryn Becker-Blease, Oregon State University, provided an overview of the field of implementation science, including how it can be applied to sexual harassment prevention efforts in institutions of higher education. Their presentation was based on a paper commissioned by the workshop planning committee.[3] The paper includes a summary of the organizational barriers to preventing sexual harassment; defines implementation science, including a comparison to other intervention research fields; outlines a subset of the research methods, designs, and models used in implementation science; and provides examples of models of implementation science that may be relevant to evaluating sexual harassment prevention efforts. The paper also includes an overview of potential barriers and next steps for approaching sexual harassment prevention from an implementation science lens.

Soicher discussed multiple outcomes of interest related to implementation, including how they could be applied to the evaluation of sexual harassment prevention, which include acceptability, adoption, appropriateness, feasibility, cost, reach, sustainability, and fidelity. (See Appendix E for worksheets that demonstrate how these outcomes of interest can be applied to the evaluation of sexual harassment prevention efforts.) In fact, fidelity, she noted, is at the forefront of implementation science. Implementation science can be used to understand implementation processes, identify contextual influences, and assess external validity. It can also be used to identify critical components of a program and describe what factors may influence successful implementation. Implementation science design methods, including those used for within-site and between-site analyses, are also discussed in the commissioned paper.

---

[3] Available: https://www.nap.edu/catalog/26279.

One of the strengths of the implementation science approach is that it can support and facilitate evaluation and be easily adapted and applied in varying contexts. Foundational to this approach, Soicher stressed, is the need to engage the community in the evaluation. Partnerships with the community and between researchers and practitioners are critical. Soicher noted that effective practices, implementation, and enabling contexts, for example, collaborating with teams, can result in improved outcomes (see Figure 3-3).

> **Implementation science** is the systematic study of issues related to the adoption, use, and generalizability of an evidence-based practice.

Soicher presented the Practical, Robust Implementation and Sustainability Model (PRISM) as it applies to Vanderbilt University's effort to alter departmental admission policies. The effort was designed to diffuse dependent relationships between graduate students and their advisors (see Box 3-1 and Figure 3-4; also see Case Study C in Appendix D). PRISM is an extension of a widely used planning and evaluation framework known as Reach, Effectiveness, Adoption, Implementation, and Maintenance (RE-AIM). Soicher noted that this model can be used to improve implementation efforts as well as develop potential intervention and implementation questions.

Becker-Blease provided an overview of the Consolidated Framework for Implementation Research (CFIR) as applied to an initiative of the

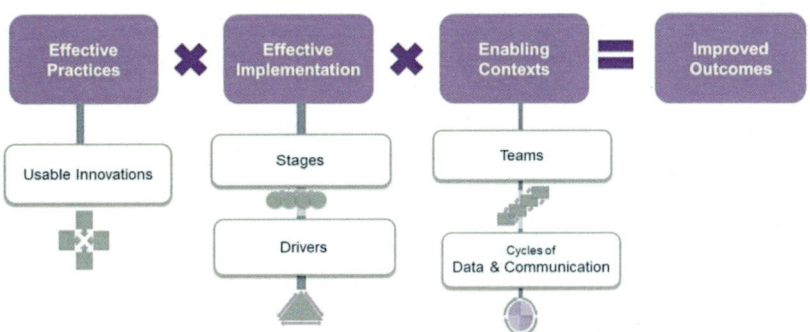

FIGURE 3-3 Formula for success and active implementation frameworks.
SOURCE: Raechel Soicher and Kathryn Becker-Blease (2021). Workshop presentation. Adapted from Fixsen et al., 2005, 2015.

Massachusetts Institute of Technology to offer lab-based inclusive culture workshops (see Figure 3-5 and Box 3-2; also see Case Study E in Appendix D). This framework synthesizes common constructs from across multiple implementation theories and provides a consistent taxonomy for building a knowledge base around what works where and why. The CFIR can outline detailed construct definitions and steps for research processes and consists of five major domains including:

- intervention characteristics, or the core components and aspects which should be preserved to maintain the effectiveness of the intervention;
- outer setting, or economic, political, and social contexts that influence implementation of and intervention within an organization;
- inner setting, or the local culture, climate, and structure of the organization which affects implementation;
- individual characteristics, which refer to recipients of the intervention and their knowledge and beliefs about the intervention; and
- process, or details of the active change process.

The key challenge, stated Becker-Blease, is assessing whether gaining buy-in from faculty will ultimately have an impact on the effectiveness of the program. The framework may be used to assess training impact for this case example. (See Chapter 6 and the associated commissioned paper[4] for a discussion of resources to support the implementation of this framework.)

---

[4] Available: https://www.nap.edu/catalog/26279.

BOX 3-1
Case Study C: Altering Departmental Admissions Policies to Diffuse Dependent Relationships Between Graduate Students and Their Advisors— Vanderbilt University

Graduate programs in biological/biomedical sciences may use any of three distinct types of admission processes. The most common is for several programs to band together and offer admission through an umbrella program in which accepted students share a common first-year curriculum, meet several potential advisors from many graduate programs, and then join an advisor's lab within a graduate program. Vanderbilt has two umbrella programs. In a second route, a graduate program may accept students into its specific program and the students can meet with several potential advisors and select one. The third process is for a student to apply to a graduate program to work with one specific advisor, chosen in advance of applying. This is a direct-admit process. Direct-admit students may have less information for choosing an advisor, and they have less power to change labs if problems arise.

Observing that direct-admit students seemed to have a more difficult path in graduate school than other students, with more advisor-student conflict and increased rates of leaving a Ph.D. program before graduation, the university decided to create a new policy for its Department of Cell and Developmental Biology (CDB), which uses all three admission policies. Under the new policy, students must apply to a Vanderbilt umbrella program to be considered for direct admission to a lab at CDB. Only application, not admission, is required. Students are expected to work in an advisor's lab before applying to that lab through the direct-admit program. If they have not, they must explain why this is not possible in their application materials. Three metrics are being used to measure success of the policy. One is the experience of the direct-admit graduate students for the next 6 years. The second is to measure how many direct-admit graduate students have previously worked with their advisor prior to arriving. An increase in this percentage will be considered a success of the policies. The third measure is the overall number of direct-admit graduate students. A decrease in their number will be considered a success of the policies.

> **BOX 3-2**
> **Case Study E: Lab-Based Inclusive Culture**
> **Workshops—Massachusetts Institute of Technology**
>
> Lab-based workshops were developed at the Massachusetts Institute of Technology (MIT) as an initiative to reach graduate students and postdocs and train them about issues and resources related to gender bias and sexual harassment. The lab serves as an important unit of community within the MIT environment and the lab workshops provide an opportunity for MIT resources to connect with the lab and help reestablish norms. The offices that have previously led this initiative approached this work at the departmental level instead of by request from individual labs to amplify the impact of a department initiative. These offices would partner with the students and department leadership, conduct a faculty demo during a regular faculty meeting, and then the department chair would put it to a vote so that this could be a faculty-driven initiative with the appropriate buy-in and engagement. Workshops are currently being implemented, including virtually. Additionally, the workshops have been evaluated through learning-based outcomes and a 6-month post-test evaluation to determine the long-term effects of the training.

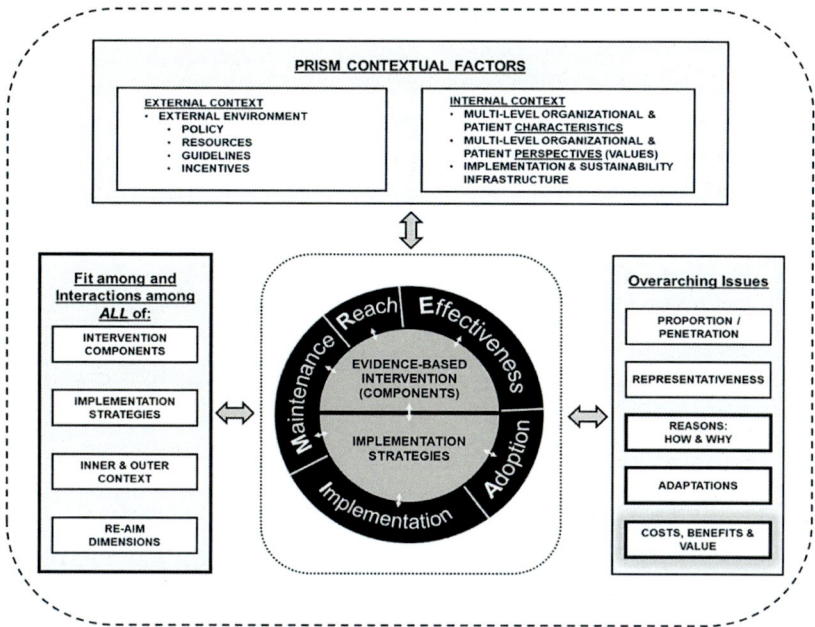

FIGURE 3-4 Practical, Robust Implementation and Sustainability Model (PRISM).
SOURCE: Raechel Soicher and Kathryn Becker-Blease (2021). Workshop presentation. Adapted from Feldstein & Glasgow 2008.

FIGURE 3-5 Consolidated framework for implementation research.
SOURCE: Raechel Soicher and Kathryn Becker-Blease (2021). Workshop presentation. Adapted from Damschroder et al., 2009, 2011.

4

Challenges and Limitations that Arise When Evaluating Sexual Harassment Prevention Efforts

Throughout the workshop, participants discussed the challenges and limitations in developing and implementing evaluation for sexual harassment prevention programs. This chapter summarizes several of the issues and barriers identified.

Several participants discussed challenges in evaluating prevention efforts that are unique to sexual harassment. Clemencia Cosentino, National Science Foundation, underscored the importance of developing a deep understanding of the unique challenges facing evaluation of sexual harassment prevention efforts to ensure the development of rigorous methods that result in useful, actionable findings. She noted that Eden King, Rice University, had stressed that selection bias, emotional reactivity, confidentiality, legal considerations, biases, and more make this work extremely challenging. Other speakers had recognized that evaluation is complicated by the fact that sexual harassment often goes unreported. Despite these issues, participants reinforced how important it is to evaluate these programs.

Kelley Bonner, National Oceanic and Atmospheric Administration, was one of several participants who noted that a lack of knowledge, time, and experience with evaluation, including survey development, pose barriers to developing and implementing evaluation programs. Robert Brinkerhoff, Western Michigan University, reiterated this point, noting that evaluation does not have to be challenging. Brinkerhoff suggested that programs just begin to collect data to answer questions about how well, for example, a sexual harassment prevention training is being used, when it is being used, and why. He added that training programs often get predictable results, for example, some participants will apply their learning outcomes, and some will not; thus, an evaluation of who has been reached can help to determine how to increase the effectiveness of the program. In essence, examining the failure of an intervention can be just as important as evaluating its success. Brinkerhoff noted, "Evaluation and program efforts can mature as you go." Kurt Kraiger, University of Memphis, stated that programs should "start by doing something."

Several participants noted that a common barrier is a lack of communication, particularly with leaders, around the need for culture change and how inclusivity can and should be nurtured. As described further in Chapter 5, Karen Stubaus, Rutgers University, said that there is a need to engage leadership on these issues, including identifying which evaluative measures to use to make the best case. Readiness for change evaluation is also a challenge, she added. She also noted that engaging leadership can be complicated by the constant change that is occurring within institutions of higher education. While change can be challenging for any organization, it is particularly difficult for institutions of higher education given that many of those in leadership positions come from faculty ranks and have not participated in leadership training, further complicating and lengthening the change process. Other participants reflected on challenges, including a lack of a long-term commitment to these efforts, continuity of evaluation and funding, and an institutional structure to support the work, noting these could be addressed with stronger leadership and support.

> People sometimes want quick terminology fixes rather than engagement goals. How do we assess and improve in a workplace that allows people to self-correct? And how do we create environments in which people can make mistakes?
>
> —Workshop participant

Several participants noted that current efforts to evaluate sexual harassment prevention efforts often do not address issues of intersectionality and identity-based discrimination. As described more fully in Chapter 5, Jennifer M. Gómez, Wayne State University, said often there is significant attention paid to developing a strong sexual harassment prevention program, while diversity, equity, and inclusion (DEI) are considered an afterthought. This is a challenge for those who are designing and implementing evaluation programs and want to ensure that the programs are reaching those most likely to be targets of harassment. Participants added that evaluating programs through an intersectional lens may surface an additional challenge—the "small N-problem," in which there are only a few people of color, as faculty or students, who are the focus of the program and evaluation but may not be able to share candidly in an anonymous way. King noted that examining intersections, particularly in smaller institutions where there are very few people of color, can make these groups further vulnerable to labeling or harassment.[1]

It was noted by several participants that limited resources serve as a barrier for institutions evaluating their sexual harassment prevention programs. To address this, organizations facing the resource challenges (including community colleges and institutions financially affected by the COVID-19 pandemic) might consider developing partnerships as part of the solution, stated one participant. Regional partnerships and partnerships across systems of higher education can be helpful to leverage resources. Participants also suggested other resources to support organizations with limited budgets for evaluation, including the National Sexual Violence Resource Center (NSVRC) evaluation toolkit[2] and the Strategic Prevention Solutions Resources.[3]

---

[1] Privacy is an important consideration when evaluating the experiences of underrepresented populations, especially people of color, sexual and gender diverse individuals, and those with disabilities, who experience higher rates of harassment than their white, cisgender, and heterosexual colleagues (Basile et al., 2016; Cortina, 2004; Lombardi et al., 2002; Settles et al., 2008).
[2] Available: https://www.nsvrc.org/prevention/evaluation-toolkit.
[3] Available: https://www.strategicpreventionsolutions.com/digitaldownloads.

5

Participant Reflections on Workshop Presentations and Discussions

Throughout the workshop, participants discussed ways to make progress in evaluating sexual harassment prevention efforts in higher education. They also discussed actions for moving the field forward, including ways to support and strengthen evaluation in sexual harassment prevention in higher education. During some of the workshop sessions, Layne Scherer, a visual practitioner and senior program officer at the National Academies, assimilated ongoing discussions and produced graphics for active ease of viewing. Figure 5-1 shows such a graphic from a session in which Eden King, Rice University, shared key themes from Day 1 of the workshop. This chapter covers participants' reflections, which are described below.

TAKING A SYSTEMATIC AND OUTCOME-FOCUSED APPROACH

Workshop conversations centered on the notion that evaluation ought to be approached systematically. There is a need to examine sets of programs and relationships in context to look at the issue holistically: Examining the system as a whole allows for clearer comparisons and the ability to generalize, King noted. This has not yet happened; despite the knowledge that programs are embedded in systems, most evaluation methods have been focused on a specific program. Armando Estrada, Temple University, added that thinking more systematically about evaluation, including on how to intervene, is vital; these are complex problems that require complex solutions. As part of this systemic approach, King, Clemencia Cosentino (National Science Foundation), Elissa Perry (Columbia University), and others noted that considering context is critical in approaching the evaluation of sexual harassment prevention in higher education. The power dynamics are unique in higher education, and this has great implications for the type of evaluation work that is needed.

Further evidence of this need for a systemic approach was supported by the frameworks presented through the prevention and implementation science lens, as discussed by Cindy Crusto, Yale University School of Medicine; Lisa Hooper, University of Iowa; and Raechel Soicher and Kathryn Becker-Blease, Oregon State University. These approaches offered lessons on how to conduct evaluation; while the approaches varied, participants made the point that evaluation should be conducted systematically and strategically.

Citing institutional climate surveys as an example of the need for a more systemic approach to evaluation, Estrada noted that while these surveys may be able to help provide baseline data, they may not be able to capture the complexity of these issues. These surveys may be more useful for diagnosing than prescribing, as noted by Brian Martinson, HealthPartners Institute.

Additionally, workshop participants reflected on the need to approach evaluation of sexual harassment prevention efforts with a focus on the intended outcomes. At the start of the workshop, Perry had stated that outcome measures should be aligned with both program purpose and time frame (short term and long term). Perry also noted that an evaluation process often unfolds over time; in the initial stages, it is often an exploratory process. However, as time goes on, evaluation can help to assess the effect of programs. Establishing an evaluation before implementing a program is important as it encourages discussion about what the program is intended to accomplish. It also provides an opportunity to determine when, how, and what data are needed.

Perry suggested that including those conducting the evaluation in program planning conversations early also allows for planning an evaluation in collaboration with the program. King had also highlighted that defining goals of the program in context is critical. The need to clearly specify outcomes is particularly important given the range of prevention programs and their varied purposes, noted King. Crusto added that it is important to recognize that no one culture or context is static; things change all the time—the outcomes that one had hoped for at one time may be affected by

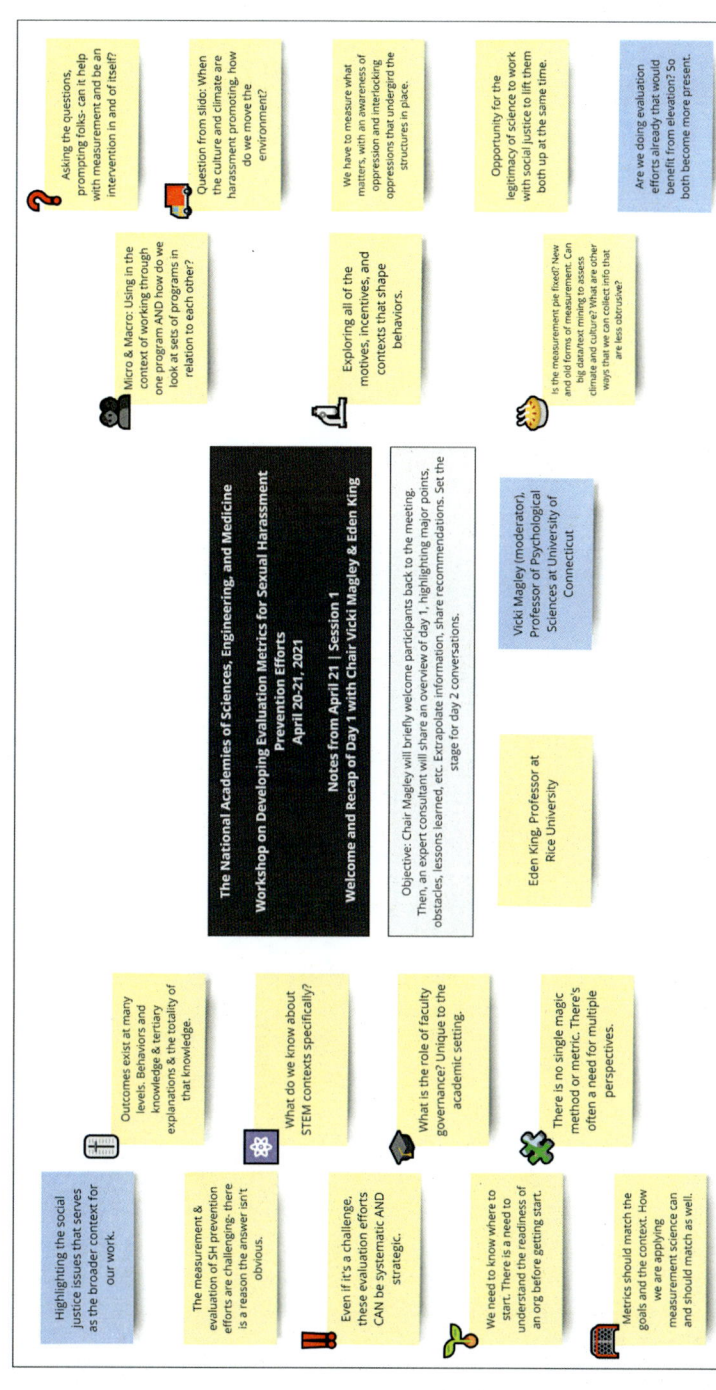

FIGURE 5-1 Summary of key themes from Day 1 of the workshop.

these changing contextual factors and, as such, one may not get the same outcomes in an ongoing or sustainable way.

King noted that outcomes exist at many levels, which complicates evaluation of sexual harassment prevention; evaluations must capture not only behavior or knowledge, but also the secondary and tertiary implications of programs and policies. She added that there is a need to broaden the notion of outcomes to be inclusive and related to the ultimate goals of the program. Relatedly, metrics should match the goal, stated King. As one participant noted, if the goal is to provide knowledge about sexual harassment through prevention efforts, one should measure that knowledge. There is also a need to measure ideas and constructs from multiple perspectives, another participant commented. King noted that needs analysis and change readiness can be used to support these efforts. Vicki Magley, University of Connecticut, noted that the distinction between how to measure effectiveness and the choice of study design is a big one. In organizations, people often ask "how do I measure . . . ," but they really are asking questions about inference, and valid inferences ensue from strong research designs.

CONSIDERING THOSE MOST VULNERABLE TO HARASSMENT

Participants noted the importance of building relationships during evaluation, centering around those marginalized groups most likely to be targets of harassment (e.g., historically excluded groups, international faculty and students, undocumented students, etc.). As one participant noted, "If you want to do evaluation, you need to build relationships first. We are embedded in a culture that does not allow diverse voices to collaborate and be present; putting a survey out is not intersectional." King noted the importance of relationships, particularly in institutions with only a few people of color in faculty who may be included in an evaluation, and highlighted that expanding efforts beyond surveys or subgroup analyses will help cultivate such relationships.

As mentioned in Chapter 4 and reinforced by Melissa Kwon, University of California, Berkeley, it is important to make the connection between evaluating sexual harassment and diversity, equity, and inclusion (DEI) explicit. Gómez, Kelley Bonner (National Oceanic and Atmospheric Administration), and others added that there is a need to carefully consider the connection between sexual harassment and DEI in evaluation, particularly as DEI efforts are needed to protect groups at highest risk of harassment. As Nelia Viveiros, University of Colorado, Anschutz Medical Campus, stated, there is a need to tie in issues of intersectionality into sexual harassment programs. Micro-interventions and proximal goals could be useful to this end and to better gauge progress. As Bonner notes, "we create this false division between sexual harassment and DEI efforts. We completely exclude the concept of intersectionality and its huge impact on prevention and response efforts for harassment."

> We have to stop checking boxes and start building systems that address identity-based discrimination.
>
> —Nelia Viveiros, University of Colorado, Anschutz Medical Campus

Complicating this challenge, said Theodosia Cook, University of Colorado, DEI leaders often find themselves as the only people with an equity lens at the leadership table; it is also very likely that they are the only woman, person of color, or immigrant. In this role, they are also too frequently discounted by people in academia. There is a need to create conditions for DEI leaders to lead, she added. As one participant noted, it is important to use an intersectional lens for both programs and measurement, and to build collaborations with others, including DEI staff and leaders.

More training on intersectionality is needed, one participant noted, because "many of us in the sexual violence field are not trained on DEI, and many in the DEI field are missing training on the layered impact of sexual violence." In fact, another participant noted, "I'm more interested in evaluation of accountability and culture change at the top of the hierarchy than training programs. Misunderstandings at the top are part of the reason DEI and sexual harassment are handled separately." Others noted that continuing to develop measures that capture power dynamics is important in this space, particularly given that these dynamics persist in higher education.

> DEI work is everyone's work. We need to trust, value, and respect the leadership of those in DEI roles. Until then, we may spin our wheels.
>
> —Theodosia Cook, University of Colorado

One participant noted: "Some of our communities that experience marginalization have also said it is embarrassing to have to keep 'justifying' why we need to do this work. It belittles their humanity."

LEADERSHIP AS A LINCHPIN: BUILDING SUPPORT AND COMMUNICATING VALUE

Workshop participants reflected on the need to build the interest and capacity of leadership to support evaluation of sexual harassment prevention efforts. Results from evaluation efforts can be used to encourage leaders to be more involved in sexual harassment prevention. Reinforcing this point, Bonner stressed the importance of communication with leaders; providing leaders with the information from evaluations, logic models, and other sources can help to develop the case for improving programs. Discussion with leaders about sexual harassment as posing a "risk" to the institution is also critical, noted Bonner. As one participant noted, "We have to figure out how to make leadership more afraid of not knowing than knowing."

Estrada added that leaders may recognize that their institution has a problem but may not have a concrete understanding about how to address it. Through evaluation, they may have a framework they can use to unpack a complex situation and translate action into research. Additionally, continuous leadership and engagement was raised as particularly important by participants discussing Case Study D (in Appendix D), which describes an effort by Argonne National Laboratory to encourage employees to recognize colleagues for demonstrating the institution's "core values." In this case study, the program was originally set to run for a short period, but through leadership's support it was extended because of its value as a positive tool for recognition and encouragement of positive behavior.

Bonner noted that it is the job of prevention staff and others managing sexual harassment prevention efforts to support leaders so that they feel

> My next step is strategic/political; convincing powerful stakeholders that systematic evaluation of existing interventions is needed to produce the outcomes they want.
>
> —Elizabeth Hutchison, University of New Mexico

confident and competent to address these challenges. Successful evaluation of sexual harassment prevention efforts requires the ability to appeal to different types of leaders through compelling arguments about the importance of these programs and efforts. As one participant noted, evaluative measures could be used to convince leadership about the value of prevention programs.

> How do we use evaluation to create system change?
>
> —Kurt Kraiger, University of Memphis

In discussions about how to create incentives to engage leadership, Karen Stubaus, Rutgers University, and others suggested developing or incorporating outcomes related to sexual harassment in U.S. News–type rankings or other rankings of best places to work. Janine Clayton, National Institutes of Health (NIH), noted that institutions should also consider opportunities to encourage adoption of proven strategies, such as NIH's Challenge Prize for Enhancing Gender Diversity, which recognizes institutions that have achieved sustained improvement in gender diversity. Participants also cited the STEMM Equity Achievement (SEA) Change program of the American Association for the Advancement of Science (AAAS), which aims to advance institutional transformation in support of diversity, equity, and inclusion in higher education institutions. Another participant noted that evaluation research that supports the idea that people can stop the harm in the first place can be used to convince leadership of the program's value. Evaluation is the basis for getting funding and demonstrating outcomes and can allow practitioners to appeal to leaders, stated Kiana Swearingen, University of Washington.

EXPANDING THE EVALUATION TOOLKIT

King and others noted the importance of considering a broader range of tools, both those that have been used historically and those that are new and innovative, in the evaluation of sexual harassment prevention efforts. This could include text mining for techniques or papers that have examined climate culture attitudes or identifying new forms of data analysis that may be less obtrusive. Pulse surveys, in which participants are asked one question per week, were also discussed by King and others. These surveys may allow for more immediate insight into measurement of changes given their frequency. King and others noted that bystander intervention research and other older techniques may be useful to apply to the measurement toolkit. Kurt Kraiger, University of Memphis, added that measures, such as a sense of psychological safety could be useful, as well as understanding the number of women (and others) who leave the institution. Another participant noted that as institutional climate is the strongest predictor of the presence of sexual harassment, it may be important to focus on metrics that focus on this predictor.

> Evaluation has to start at the very beginning of the project.
>
> —Jane Stapleton, Soteria Solutions

Participants discussed the importance of incorporating evaluation from the beginning. Michael Leiter, Acadia University and Deakin University, noted that talking to people within an organization about what problems they are facing may be a good place to start. "It is also critical to think outside the box about what evaluation can look like at the outset, including to the target audience," stated Jane Stapleton, Soteria Solutions. The target audience is central to understanding what needs to be evaluated and should be involved in identifying new solutions and methodologies, as noted by one participant. As Kwon discussed, a program should be developed alongside stakeholders from the beginning and throughout.

Related to conversations about building trust, multiple presenters discussed approaches for improving participant engagement. Leiter stated that strong participant engagement is crucial to the program success and the ability to collect accurate data. This is particularly true in the health care industry where staff are often over-surveyed. The risk of over-surveying participants is particularly important to consider for marginalized populations; we need to use these surveys and data responsibly, stated Swearingen. There is also a need for larger discussion at the institutional level about what we are asking of those we are surveying, stated Stapleton, adding that there are other ways to collect data than surveys.

BUILDING COMMUNITY, COLLABORATION, AND TRUST

Participants reflected on the importance of continuing to collaborate and share resources, working together to support evaluation efforts in higher education. Swearingen and others discussed the need to continue to share resources and work together to move toward broader evaluation of sexual harassment prevention efforts in higher education. This may mean moving away from the mindset that sharing information about available resources can be detrimental in situations where there is a scarcity of those resources. Swearingen offered an example of an unhelpful scarcity mindset: "If I do something successful, I'm not going to share it, because then I might not get funding." The idea of sharing information requires a fundamental culture change for academia.

Participants reflected on the importance of transparency and trust in the population being evaluated. Kwon and others reiterated the importance of transparency and trust in developing and evaluating programs designed to prevent sexual harassment. Building trust among participants is critical and must be developed by engaging participants throughout the process, noted several participants. As Stapleton said, providing participants with data throughout the evaluation process, including after the work is completed, can help build trust and transparency, which are particularly important in issues related to sexual harassment.

> We need a space to continue to build trust and be in long-term relationships with others to make change.
>
> —Kiana Swearingen, University of Washington

6

Evaluation in Action: Examples and Resources

Many higher education institutions are actively working to evaluate their efforts to address sexual harassment, but those evaluations may or may not be consistent with what research has identified as best practices. To this end, the workshop included three examples of organizations that have had success with aspects of evaluating their interventions. Additionally, this chapter offers resources for applying implementation science to the evaluation of sexual harassment prevention efforts.

EXAMPLES OF EFFORTS TO EVALUATE SEXUAL HARASSMENT INTERVENTIONS

Michael Leiter, Acadia University and Deakin University, provided the first of three examples of evidence-based efforts to evaluate sexual harassment interventions. He discussed his research on an initiative designed to promote and evaluate workplace civility, including how employees show respect and disrespect at work, called strengthening a culture of respect and engagement (SCORE). This initiative, which has been implemented in various countries, including in Australia and North America, is held in five sessions, each 90 minutes in length, over 2 to 4 weeks apart; it assesses civility, incivility, and levels of institutional intimidation among employees. SCORE also includes a facilitated group process to strengthen social climate in the workplace. It has been used in a variety of workplaces and is based on the components of the CREW (civility, respect, and engagement at work) initiative, which has been deployed in veterans hospitals across the United States.

Social engagement scales were created to assess interactions with ratings from civil to uncivil. The Maslach Burnout Inventory, measured on three dimensions—exhaustion, cynicism, and inefficacy—assesses employees on scales ranging from engaged to burnout. The evaluation design also included treatment and control groups. Over time, based on results from various workplaces, the program resulted in higher sources of civility among coworkers and individuals, while sources of incivility also decreased among supervisors (see Figure 6-1).

To evaluate impact, Leiter noted that he and his colleagues relied on diverse sources of information, including surveys, institutional data, interviews, and observations, including validated measures to assess experience at work and social dynamics. The evaluation design included before and after assessment and comparison groups.

Following Leiter's presentation, Jane Stapleton, Soteria Solutions, discussed the Workplace Violence Prevention and Response Program Collaboration of the National Oceanic and Atmospheric Administration (NOAA), designed to identify sexual assault and sexual harassment (SASH) protective and risk factors for different NOAA workplace environments. The effort was designed to determine social norms, recognition, bystander behaviors, and prevalence of SASH and other inappropriate behaviors. It was also developed to support NOAA-specific strategies that are knowledge- and skills-based and tailored to NOAA workplace environments. The current evaluation uses public health approaches to prevention, based on data to define the problem, inform solutions, make course corrections, and measure outcomes. The evaluation begins with the start of the project, involves multiple methods, actively engages the target audience, and uses a variety of indicators to measure outcomes. The evaluation cycle begins with a needs assessment, followed by a workplace culture survey and a formative evaluation, leading to piloting the prevention strategy. This work is followed by course correction and refinement strategies, a program evaluation, and, finally, outcome evaluation.

To develop the needs assessment, Stapleton explained, focus groups and interviews with NOAA employees were conducted, along with examinations of existing data on SASH complaints and investigations. A review of current training was also conducted. The needs assessment yielded several learning outcomes, including how variable the work environments, cultures, and readiness levels are across the agency. Stapleton added that

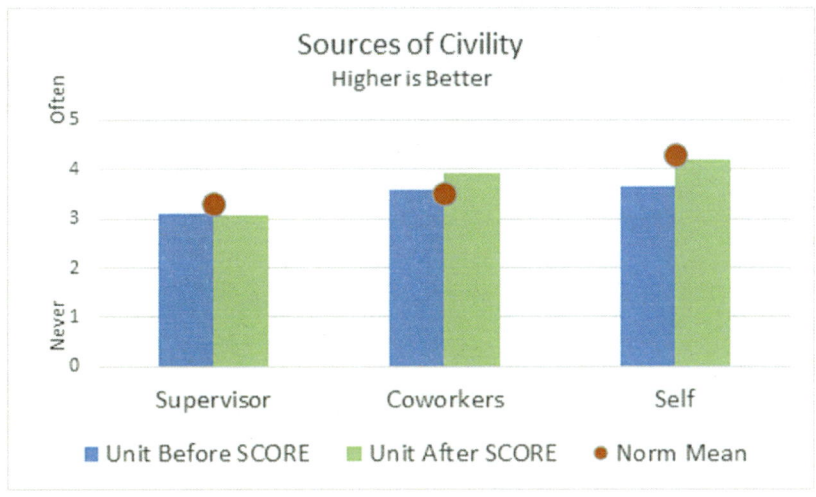

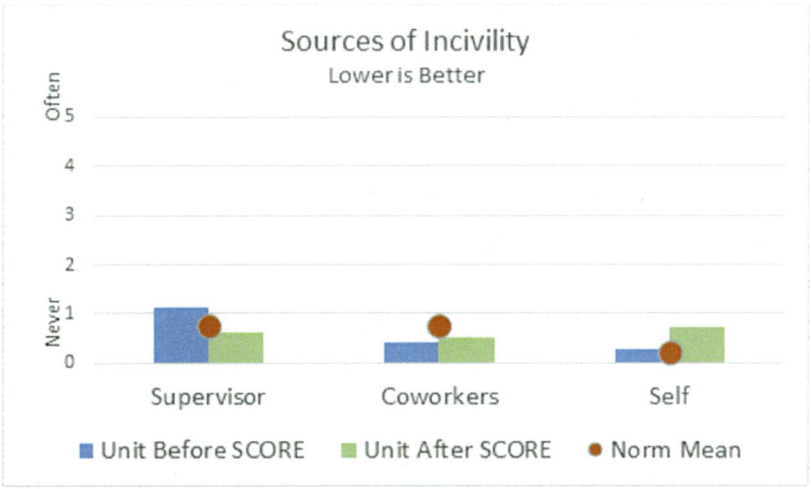

FIGURE 6-1 SCORE evaluation: Sources of civility or incivility.
SOURCE: Michael Leiter (2021). Workshop presentation.

the assessment reinforced the need for consistent messaging throughout the organization and core training, as well as the importance of leveraging social norms to change behaviors.

In terms of program evaluation, Stapleton noted that the key interest is around whether the program is meeting its objectives. To determine if the program is being implemented as designed, the investigators are planning to track participation and assess how the program implementation is functioning from administrative, organizational, and personnel perspectives.

In terms of outcome evaluation, the focus will be on assessing the:

- increase in recognition of SASH and other inappropriate behaviors;
- increase in bystander, de-escalation, and strategic resistance behaviors;
- increase in positive social norms that support safe and respectful workplace environments;
- increase in reporting of SASH and other inappropriate behaviors; and
- decrease in SASH and other inappropriate behaviors at NOAA.

Qualitative methods have been important to this work, stated Stapleton. As one participant noted, qualitative methods can delve into intersectional experiences, process, and power dynamics, and highlight what is going on in ways that are sometimes hard to capture with surveys.

Melissa Kwon, University of California, Berkeley, discussed the university's PATH to Care Center program, which includes sexual harassment prevention initiatives, advocacy, training, and healing services. The cornerstone of the program is its population-specific prevention managers who work directly with undergraduate students, graduate students, staff, and faculty.

Kwon discussed the program's evaluation process, which includes determining short-, intermediate-, and long-term outcomes of prevention programs and survivor support programs; developing goals of training and other programs; examining existing evaluation questions; mapping existing evaluation questions onto outcomes; identifying indicators for each outcome; developing new survey questions to fill gaps; designing new interview and focus group questions; and creating a "menu" of evaluation questions. The evaluation "menu" allows for individual program managers to pull from a list of questions to support their evaluation (see Figure 6-2). The program also captures information through a service tracking form.

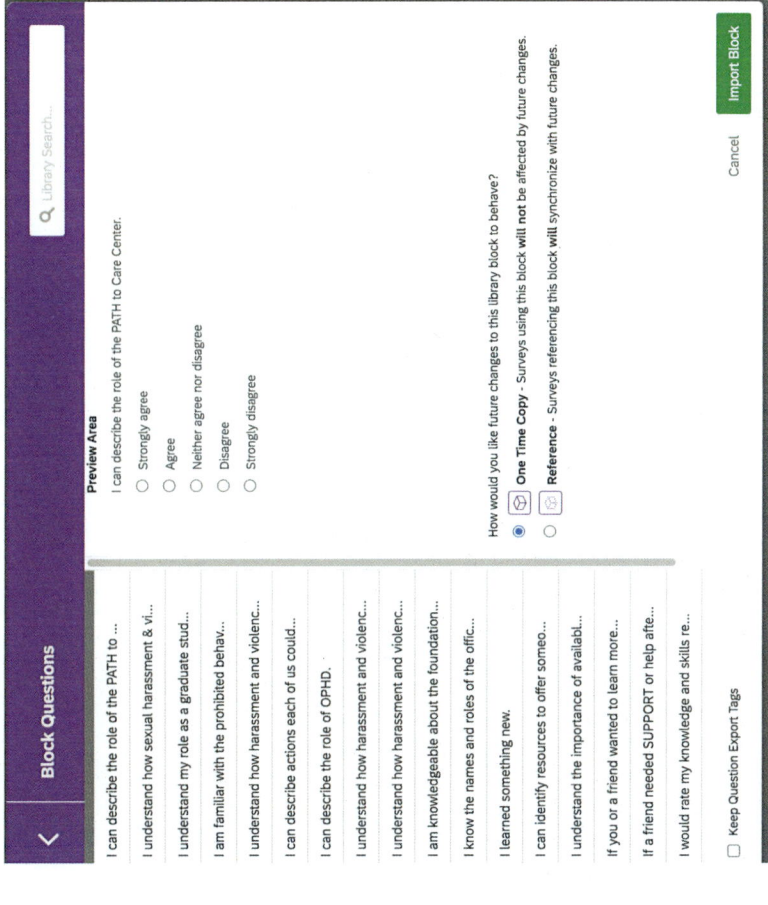

FIGURE 6-2 Example of an evaluation "menu."
SOURCE: Melissa Kwon (2021). Workshop presentation.

Kwon noted that the staff are in the process of refining their data collection systems, continuing their routine data collection, demonstrating to the team how data are being used, determining how leadership and the team want to use the evaluation data, and assessing what data will be included in a shared dashboard of information.

APPLYING IMPLEMENTATION SCIENCE TO UNDERSTAND CONTEXTUAL FACTORS THAT FACILITATE OR IMPEDE SUCCESSFUL EFFORTS

In addition to the efforts to evaluate sexual harassment initiatives described above, there are also several examples of evidence-based frameworks that can be applied to the evaluation of sexual harassment prevention efforts. The commissioned paper by Raechel Soicher, Oregon State University,[1] provides several such examples, including resources that outline how to apply models of implementation science to the evaluation of sexual harassment prevention efforts. In addition, Appendix E includes resources and worksheets based on these models.

As described by Soicher and Kathryn Becker-Blease, Oregon State University, the first step in conducting an implementation study is selecting a model, which can help at all phases of a research study, from planning to execution to evaluation. As described in Chapter 3, one resource that can help facilitate this process is the use of a logic model. Figure E-1 in Appendix E offers a logic model of an implementation science project that could be adapted to a sexual harassment prevention program. The model begins with an assessment of starting conditions, in which one outlines the problem or gap; continues with an analysis of inputs and proximal outcomes; and concludes with the identification of distal outcomes, including the long-term outcomes of the project. Figure E-2 includes a logic model worksheet to support this analysis, offering options for a user to identify what factors are relevant to the project, particularly as related to the problem; the program, intervention, or policy and implementation strategies; and outcomes.

The Consolidated Framework for Implementation Research (CFIR), as described in Chapter 3, synthesizes constructs from across multiple implementation theories and provides a consistent taxonomy for building a knowledge base. The framework is commonly cited in the literature, as

---

[1] Available: https://www.nap.edu/catalog/26279.

presented by Becker-Blease, and can be effectively applied to the evaluation of sexual harassment prevention efforts. Table E-1 (in Appendix E) includes a worksheet to support the application of the CFIR to sexual harassment prevention, including offering guiding questions for a user to assess intervention characteristics; outer-setting characteristics (e.g., stakeholders' needs and resources, external policies and incentives); inner-setting characteristics (e.g., structural characteristics, culture, implementation climate, readiness for implementation); characteristics of individuals; and process issues (e.g., planning and engaging).

Also as described in Chapter 3, the Practical, Robust Implementation and Sustainability Model (PRISM) can improve implementation efforts, as well as helping to develop potential intervention and implementation questions; it can be applied to the evaluation of sexual harassment prevention efforts. Table E-2 (in Appendix E) includes a worksheet to guide the application of this model to sexual harassment prevention efforts. The worksheet includes guiding questions to help a user describe and assess the program intervention, environment, implementation and sustainability infrastructure, and program recipients at the organizational and consumer levels.

These resources are select examples of implementation science frameworks and models that can be used as a guide for higher education institutions in applying implementation science to evaluation of sexual harassment prevention efforts. Additional information about these and other models can be found in the commissioned paper by Soicher.[2]

---

[2] Available: https://www.nap.edu/catalog/26279.

7

Reflections by Eden King on Themes and Next Steps

The analysis in this chapter was commissioned by the planning committee for this workshop. Eden King, Rice University, is an expert in the field and was tasked to synthesize key ideas, connections, and themes shared at the meeting, as well as to suggest actionable next steps to "move the needle" on the evaluation of sexual harassment prevention efforts. The opinions expressed in this chapter reflect the views of Eden King and are written in first person to retain her voice. This shift to a first-person narrative distinguishes her paper from the third-person summaries of workshop discussions in the previous chapters.

THE UNIQUE CHALLENGE

Measurement and evaluation scholars and practitioners deal with several challenges (see American Educational Research Association [2014] for a review of some of these issues). What is unique about measurement and evaluation in the context of efforts to address sexual harassment is that *all* of the challenges apply. In other words, *everything* that makes good measurement and evaluation difficult is amplified in the context of sexual harassment.

The challenge of conducting high-quality evaluation of sexual harassment prevention efforts includes such issues as selection bias, social desirability concerns, construct ambiguity, event history, consequential validity, emotional reactivity, confidentiality, legal considerations, introspective biases, retrospective biases, and more. At the same time, evaluation plays an essential role in the extent to which any program can be implemented effectively and sustained consistently over time. This tension between the challenges and the essential nature of high-quality evaluation in harassment prevention served as the impetus for a workshop. This paper briefly synthesizes my perspective on some of the major themes that emerged in the presentations and discussions during the workshop, as well as my own personal reflections and suggestions for moving forward.

THEMATIC SYNTHESIS

Theme 1: Evaluation as a systematic, strategic process. Workshop presenters described a variety of best practices in training and program evaluation that might be applied to the specific case of sexual harassment prevention efforts. These descriptions included frameworks developed in intervention[1] and implementation science,[2] as well as more frameworks that are typically applied to organizational training evaluation (Salas et al., 2012). Each of these disciplinary lenses offers unique emphases, but they also converge on a critical, core common principle: Evaluation can and should be a systematic and strategic process. The specific steps or generative questions vary from one framework to another (or from one disciplinary lens to another), but the overarching commonality is that evaluation will be most defendable and informative to the extent that it is based on a systematic process.

Theme 2: This process must begin at the end. Workshop presenters and attendees also converged in articulating that the systematic process of evaluation of sexual harassment prevention efforts must begin with a focus on the intended outcomes of the efforts. In other words, what are the goals of the effort itself? These might be individual-level goals (determined via an individual needs analysis or a change readiness assessment) or unit-level goals (determined via an organizational needs analysis or assessment; see Salas et al., 2012), or both. Clearly thinking through and specifying in writing what programs are intended to achieve is a critical first step in any systematic program evaluation.

Theme 3: Metrics must align with goals. It is fairly straightforward to communicate that the content of program evaluation should match the

---

[1] Available: https://www.nap.edu/catalog/26279.
[2] Available: https://www.nap.edu/catalog/26279.

goals of the program—in other words, that we should measure what we set out to change. What is less obvious, but nonetheless equally important, is that the measurement *method* should also align with the goals of the program. If the goal of a particular intervention is to increase knowledge about sexual harassment policies, for example, evaluators should assess precisely that knowledge (i.e., the goal *content*). The *method* of the assessment also needs to align with the goal and content—knowledge can be assessed in a direct manner through a declarative knowledge test. As another example, if the goal of an intervention is to improve the everyday lived experiences of women in a cell biology department, a test would not be an effective methodological approach. Instead, interviews or focus groups may be a method that would allow for deeper analysis of those everyday experiences of a small number of women. Metrics should match goals with regard to content *and* method.

Theme 4: Outcomes exist at many levels and in many forms. One of the features of sexual harassment prevention efforts is that they are often designed with multiple explicit purposes—they set out to affect not only short-term, proximal, or immediate goals (like increasing knowledge and awareness), but also longer-term, distal, or systemic goals (like improving the recruitment and retention of female faculty in male-dominated fields). These outcomes might relate to changes in cognitions, emotions, behaviors, policies, practices, or demographic patterns. Intentional and unintentional outcomes might also vary with regard to the expected level of influence—individual, unit, or institutional. This variety of possibilities reinforces the need to clearly specify and measure the goals and expected outcomes of a given program or set of programs.

Theme 5: There is no one, single best method or measure. Workshop attendees discussed a variety of existing tools to assess institutional climates (e.g., masculinity contest culture; Berdahl et al., 2018) and individual experiences (e.g., psychological safety; Edmonson, 1999). Case studies included examples of both qualitative and quantitative approaches. An implicit assumption that needs to be explicit is that, unlike other fields of science or medicine, *there is no one, single "gold standard" assessment* tool for sexual harassment prevention efforts. And importantly, this lack of a gold standard tool is not because tools haven't been designed, but because anti-harassment programs serve so many different goals (at so many different levels, see above) that no single measure or metric could possibly be sufficient. This is a profoundly important conclusion that is often misunderstood.

Theme 6: Equity cannot be ignored. Substantial discussion emerged in the workshop (in both the verbal discussions and the synchronous text chats) at the prompting of attendees regarding the intersections between harassment (and harassment prevention efforts) and equity (and diversity, equity, and inclusion efforts). One key aspect of this discussion is acknowledging and addressing the notion that our methods might be constrained by the nature of privilege and the ways it perpetuates the status quo. For example, as one workshop participant queried, "When the culture and climate is harassment-promoting, how do we get buy-in, institutional support, etc., to be able to implement and evaluate programs over time?" Who does and who does not get asked what questions and when and how and by whom has been inextricably intertwined with identity, status, and power. Evaluation conclusions must be confronted in light of the reality of inequity in our institutions.

REFLECTIONS FOR MOVING FORWARD

Context matters. Nearly every presenter in the workshop mentioned the importance of context, and indeed, it is a critical component of every model that could be applied to program evaluation. In the case of harassment prevention, *Sexual Harassment of Women: Climate, Culture, and Consequences in Academic Sciences, Engineering, and Medicine* (NASEM, 2018; hereafter, National Academies), identified some specific aspects of STEM (science, technology, engineering, and mathematics) and higher education contexts that may exacerbate harassment and increase the complexity of its prevention. These STEM/higher education–specific elements of context should not be overlooked when it comes to evaluation. For example, evaluators must consider how the embedded hierarchies and dependencies of academic life (e.g., the precariousness of academic life pre- or without tenure, principal investigators' funding of students) influence the ways that we can and should measure outcomes. As another example, the unique role of faculty governance may play into the evaluation process. It would also be helpful to clarify the ways that a nonprofit orientation influences leadership behaviors or motivations. The role of the scientific process itself, for example, in when and how work gets done (e.g., in the field or in the lab) can also influence who, how, when, and where evaluation is needed in a manner that is unique to the context of STEM in higher education.

Leverage varied methods. The tried-and-true (traditional) methodological approaches that are brought to bear on typical program evaluation processes (such as attitude or climate surveys, knowledge tests, focus groups, and interviews) have great value in the evaluation process for harassment prevention efforts. Indeed, *triangulating* across these methods (i.e., looking for consistent themes from multiple methodological lenses) is a particularly compelling way to draw robust conclusions. In addition, I think there is opportunity to utilize a more diverse set of methodological tools. New and underutilized approaches, such as text mining, experience sampling or pulse surveys, and unobtrusive observation, may provide helpful information and overcome some of the methodological challenges that are common in other approaches. These techniques could be useful complements to more traditional approaches.

Harassment prevention and diversity, equity, inclusion (DEI) are inextricably intertwined. Though different individuals, offices, and efforts are often directed toward harassment prevention and what has been labeled DEI, these topics are inherently interrelated. Efforts to prevent harassment, for example, should attend to the groups of people who are most likely to perpetrate or be targets of harassment. As another example, inclusion and equity may be key elements of organizational contexts that reduce harassment. The treatment of these efforts as distinct has set up a false separation that could ultimately make it harder to achieve the goals of each set of programs. It is likely that many of the goals of these programs are overlapping (e.g., increasing the representation of white women and women of color in STEM). From an evaluation standpoint, then, measurement may be more efficient if redundancies are explicitly addressed.

NEXT STEPS

The observations noted above and in the workshop highlight the challenging variety of requirements for high-quality evaluation of harassment prevention efforts. In my view, it would not be possible to specify one single strategy or approach to this process that could adequately inform effective evaluation across the wide range of contexts and goals. The National Academies report (NASEM, 2018) on sexual harassment and the ongoing work of the National Academies Action Collaborative on Preventing Sexual Harassment in Higher Education (see Chapter 1) yielded a helpful rubric [see Appendix D] that outlines some goals for these efforts that might serve as a starting place. Re-envisioning this checklist, I think it would be helpful to develop and communicate guiding questions or principles to evaluate harassment prevention efforts.

That is, rather than specifying exactly what measures or metrics to apply and how in every case, it would be helpful to explicitly enumerate a set of core elements of high-quality evaluations of harassment prevention programs. Some workshop attendees commented that it might be motivating to create a "scorecard" or ranking systems through which institutions could strive for external recognition. While there may be value in this kind of model, I think it might be especially valuable to identify the key components of evaluation approaches that can be customized to the needs of a particular institution with particular goals. As a potential example, a guiding principle might be that evaluations should be a systematic, strategic process. As another example, it could be suggested that sexual harassment prevention effort evaluation processes should center the perspectives of marginalized members of the community. Describing such conceptual principles, guiding questions, or considerations that would clarify who should be assessed, when, in what way, and under what conditions could provide a more balanced and potentially instructive step for moving forward.

In addition to providing this guidance, I think an exciting next step for the National Academies' community would be to ideate, encourage, implement, and evaluate efforts in research and practice that explore the integration of DEI and harassment. I understand that a National Academies workshop on DEI is being organized as I type this paper; I would strongly encourage the explicit consideration of sexual harassment prevention (and the typical elements of sexual harassment prevention practice and research) in this conversation. These research programs and practices have been driven by different disciplines and interest groups, thereby artificially perpetuating their separation and limiting their development. Indeed, I think sexual harassment and DEI scholars and practitioners have the opportunity to find and build new synergies across intervention implementation and evaluation.

In sum, the workshop reflected a thoughtful conversation with great insights about the challenges of effective evaluation of sexual harassment prevention efforts and strategies through which these challenges might be met. I hope that these reflections are helpful in propelling this critical work forward.

References

American Educational Research Association. (2014). *Standards for Educational and Psychological Testing*. Washington, DC: American Educational Research Association, American Psychological Association, and National Council on Measurement in Education.

Andersson, L.M., and C.M. Pearson. (1999). Effect of tit for tat? The spiraling in the workplace incivility. *Academy of Management Review, 24*(3), 452–471. doi: 10.2307/259136.

Basile, K.C., M.J. Breiding, and S.G. Smith. (2016). Disability and risk of recent sexual violence in the United States. *American Journal of Public Health, 106*(5), 928–933.

Berdahl, J.L., M. Cooper, P. Glick, R.W. Livingston, and J.C. Williams. (2018). Work as a masculinity contest. *Journal of Social Issues, 74*(3), 422–448.

Cortina, L. M., S. Swan, L.F. Fitzgerald, and C. Waldo. (1998). Sexual harassment and assault. *Psychology of Women Quarterly, 22*(3), 419–441.

Cortina, L.M. (2004). Hispanic perspectives on sexual harassment and social support. *Personality and Social Psychology Bulletin, 30*(5), 570–584.

Cortina, L.M., M.S. Hershcovis, and K.B.H. Clancy. (2021). The embodiment of insult: A theory of biobehavioral response to workplace incivility. *Journal of Management.* doi: 10.1177/0149206321989798.

Damschroder, L.J., D.C. Aron, R.E. Keith, S.R. Kirsh, J.A. Alexander, and J.C. Lowery. (2009). Fostering implementation of health services research findings into practice: A consolidated framework for advancing implementation science. *Implementation Science 4*(1), 50. https://doi.org/10.1186/1748-5908-4-50.

Damschroder, L.J., D.E. Goodrich, C.H. Robinson, C.E. Fletcher, and J.C. Lowery. (2011). A systematic exploration of differences in contextual factors related to implementing the MOVE! weight management program in VA: A mixed methods study. *BMC Health Services Research, 11,* 248. https://doi.org/10.1186/1472-6963-11-248.

Edmondson, A. (1999). Psychological safety and learning behavior in work teams. *Administrative Science Quarterly, 44*(2), 350–383.

Feldstein, A.C., and R.E. Glasgow. (2008). A Practical, Robust Implementation and Sustainability Model (PRISM) for integrating research findings into practice. *The Joint Commission Journal on Quality and Patient Safety 34*(4), 228–243.

Fixsen, D.L., S.F. Naoom, K.A. Blase, R.M. Friedman, and F. Wallace. (2005). Implementation Research: A Synthesis of the Literature. Tampa, FL: University of South Florida, Louis de la Parte Florida Mental Health Institute, The National Implementation Research Network. Available: https://nirn.fpg.unc.edu/resources/implementation-research-synthesis-literature.

Fixsen, D., K. Blase, A. Metz, and M. Van Dyke. (2015). Implementation science. *International Encyclopedia of the Social and Behavioral Sciences, 11*, 695–702.

Glick, P., J.L. Berdahl, and N.M. Alonso. (2018), Development and validation of the masculinity contest culture scale, *Journal of Social Issues, 74*(3), 449–476.

Huang, Y. H., J. Lee, Z. Chen, M. Perry, J.H. Cheung, and M. Wang. (2017). An item-response theory approach to safety climate measurement: The Liberty Mutual Safety Climate Short Scales. *Accident Analysis & Prevention, 103*, 96–104.

Huerta, M., L.M. Cortina, J.S. Pang, C.M. Torges, and V.J. Magley. (2006). Sex and power in the academy: Modeling sexual harassment in the lives of college women. *Personality and Social Psychology Bulletin, 32*(5), 616–628.

Lombardi, E.L., R.A. Wilchins, D. Priesing, and D. Malouf. (2002). Gender violence: Transgender experiences with violence and discrimination. *Journal of Homosexuality, 42*(1), 89–101.

McCreight, M.S., B.A. Rabin, R.E. Glasgow, R.A. Ayele, C.A. Leonard, H.M. Gilmartin, J.W. Frank, P.L. Hess, R.E. Burke, and C.T. Battaglia. (2019). Using the Practical, Robust Implementation and Sustainability Model (PRISM) to qualitatively assess multilevel contextual factors to help plan, implement, evaluate, and disseminate health services programs. *Translational Behavioral Medicine 9*(6), 1002–1011.

NASEM (National Academies of Sciences, Engineering, and Medicine). (2018). *Sexual Harassment of Women: Climate, Culture, and Consequences in Academic Sciences, Engineering, and Medicine*. Washington, DC: The National Academies Press. doi: 10.17226/24994.

NASEM. (2020a). *Action Collaborative on Preventing Sexual Harassment in Higher Education: Year One Annual Report of Member Activities*. Washington, DC: The National Academies Press.

NASEM. (2020b). *Understanding the Well-Being of LGBTQI+ Populations*. Washington, DC: The National Academies Press.

Plested, B.A., P. Jumper-Thurman, and R.W. Edwards. (2006). Community Readiness: Advancing Suicide Prevention in Native Communities (Community Readiness Model Handbook). Fort Collins, CO: Center for Applied Studies in American Ethnicity, Colorado State University.

Rabin, B.A., R.G. Tabak, R.E. Glasgow, R.C. Brownson, and B. Ford. (2019). *Dissemination & Implementation Models in Health Research & Practice: A Workshop on How to Plan for, Select, Combine, Adapt, Use, and Measure Dissemination and Implementation Models in Health*. Available: https://medschool.cuanschutz.edu/docs/libraries provider94/default-document-library/di-models-workshop.pdf?sfvrsn=bb5929b9_2.

Rosenthal, M.N., A.M. Smidt, and J.J. Freyd. (2016). Still second class: Sexual harassment of graduate students. *Psychology of Women Quarterly, 40*(3), 364–377.

Salas, E., S.I. Tannenbaum, K. Kraiger, and K.A. Smith-Jentsch. (2012). The science of training and development in organizations: What matters in practice? *Psychological Science in the Public Interest, 13*(2), 74–101.

Settles, I.H., J.S. Pratt-Hyatt, and N.T. Buchanan. (2008). Through the lens of race: Black and white women's perceptions of womanhood. *Psychology of Women Quarterly, 32*(4), 454–468.

Smith, C.P., and J.J. Freyd. (2013). Dangerous safe havens: Institutional betrayal exacerbates sexual trauma. *Journal of Traumatic Stress, 26*(1), 119–124.

Substance Abuse and Mental Health Services Administration. (2019). A Guide to SAMHSA's Strategic Prevention Framework. Rockville, MD: Center for Substance Abuse Prevention.

World Health Organization. (2021). *Health Equity*. Available: https://www.who.int/health-topics/health-equity.

Appendix A

Workshop Agenda

DAY 1 (APRIL 20)

11-11:25am   **Welcome, Orientation**

This session will orient participants to the workshop's goals and outcomes, as well as to the technology that will be used to complement the meeting's discussions.

- **Vicki Magley** (*Planning Committee Chair*), Professor of Psychological Sciences at University of Connecticut
- **Arielle Baker**, Program Officer at the National Academies of Sciences, Engineering, and Medicine

11:25am-12:15pm   **Session 1: The Evaluation of Sexual Harassment Prevention Efforts in Higher Education Institutions**

This session will introduce to what is known on how higher education institutions are approaching sexual harassment prevention and evaluation, the reasons why an institution would and/or should evaluate its prevention efforts, barriers to evaluation, and evidence-based suggestions for the evaluation process. Following a brief presentation, a stakeholder panel will share their reflections and engage in a conversation with the broader group.

- **Elissa Perry** (presenter), Professor of Psychology and Education at Columbia University
- **Kelley Bonner** (panelist), Director of Workplace Violence Prevention & Response Office at the National Oceanic and Atmospheric Administration
- **Armando Estrada** (panelist), Associate Professor of Policy, Organizational & Leadership Studies at Temple University
- **Vicki Magley** (moderator), Professor of Psychological Sciences at University of Connecticut

12:15-12:45pm **Break**

12:45-1:45pm **Session 2: Measures and Metrics**

This session will focus on existing measures and predictors of organizational outcomes that might be used in sexual harassment prevention evaluation to understand their effectiveness.

- **Emily Huang**, Associate Professor at Oregon Health & Science University
- **Jennifer Berdahl**, Professor at University of British Columbia
- **Alec Smidt**, Postdoctoral Associate at Yale University School of Medicine
- **NiCole Buchanan** (moderator), Professor at Michigan State University

Following three "flash" presentations, workshop participants will be broken out into their case study subgroups to discuss and brainstorm how measures and metrics might be applied to that specific example, and to consider what other measures and metrics could be useful for this case study.

1:45-2:15pm	**Break**

2:15-3:15pm	**Session 3: Applicable Frameworks**

This session will explore basic principles of implementation science and prevention science and how they might be used to more effectively carry out and evaluate prevention efforts.

Session 3A: Implementation Science

This session will highlight how implementation science, which has been used primarily within medical and public health contexts, might be used in higher education institutions to bridge the research-practice gap and facilitate data-driven improvement of prevention efforts.

- **Raechel Soicher**, Instructor in the School of Psychological Science at Oregon State University
- **Kathryn Becker-Blease**, Associate Professor and Director in the School of Psychological Science at Oregon State University
- **Jeena Thomas** (moderator), Program Officer at the National Academies of Sciences, Engineering, and Medicine

Session 3B: Prevention Science

This session will explore how prevention science has been, is currently, and could be applied to sexual harassment evaluation in higher education, including how institutions might determine appropriate outcomes and goals.

- **Cindy Crusto**, Professor at Yale University
- **Lisa Hooper**, Professor and Richard O. Jacobson Endowed Chair for Research at the University of Northern Iowa
- **Jeena Thomas** (moderator), Program Officer at the National Academies of Sciences, Engineering, and Medicine

3:15-3:30pm **Break**

3:30-4:00pm **Session 4: Frameworks in Practice**

Presenters from Session 3 will walk participants through how implementation science and prevention science might be applied to some of the specific case studies being considered in the meeting.

- **Raechel Soicher**, Instructor in the School of Psychological Science at Oregon State University
- **Kathryn Becker-Blease**, Associate Professor and Director in the School of Psychological Science at Oregon State University
- **Cindy Crusto**, Professor at Yale University
- **Lisa Hooper**, Professor and Richard O. Jacobson Endowed Chair for Research at the University of Northern Iowa
- **Arielle Baker** (moderator), Program Officer at the National Academies of Sciences, Engineering, and Medicine

4:00-4:20pm **Session 5: Discussing Frameworks**

Participants will re-enter their case study subgroups to discuss their reactions to the framework presentations, consider how they might be applied in the context of their case study, and recognize limitations that may arise.

4:20-4:55pm **Session 6: Small Group Report-out**

Returning back to the main session, each case study subgroup will share the highlights of their discussions over the course of the day.

- **Carol Greider** (facilitator), Distinguished Professor at the University of California, Santa Cruz

4:55-5pm **Adjourn**

National Academies' staff will close the meeting.

- **Arielle Baker**, Program Officer at the National Academies of Sciences, Engineering, and Medicine

DAY 2 (APRIL 21)

11-11:30am — **Welcome and Recap of Day 1**

Following a brief welcome, this session will set the stage for Day 2 by way of an expert synthesis of Day 1 conversations. This will involve a synthesis of the ideas shared by participants thus far, an exploration of emerging connections and themes across sessions, a summary of what is still unknown and/or undiscussed, and reflections on next steps.

- **Eden King**, Professor of Industrial-Organizational Psychology at Rice University
- **Vicki Magley** (moderator), Professor of Psychological Sciences at University of Connecticut

11:30am-12:30pm — **Session 1A: Evaluation in Practice: Considerations and Limitations**

This session is a candid conversation between key stakeholders (evaluation experts, sexual harassment experts, practitioners, and institutional leaders) around their experiences with carrying out evaluation work, while considering structural barriers and factors limiting our shared advancement in this space. A major goal of this session will be to identify "sticking points" in the evaluation of interventions and engage in a group conversation (both among the panelists and all workshop participants) about how to overcome those barriers.

- **Bianca Kaushal-Carter**, Manager of Prevention, Education, and Outreach at Massachusetts Institute of Technology
- **Karen Stubaus**, Vice President for Academic Affairs at Rutgers University

- **Jennifer M. Gómez**, Assistant Professor in the Department of Psychology and the Merrill Palmer Skillman Institute for Child & Family Development at Wayne State University
- **Nelia Viveiros**, Associate Vice Chancellor for Academic Operations at University of Colorado, Anschutz Medical Campus
- **Kurt Kraiger**, Professor of Management at University of Memphis
- **Robert Brinkerhoff**, Professor Emeritus at Western Michigan University, Director of Research & Evaluation at Promote International
- **Maria Lund Dahlberg** (moderator), Senior Program Officer at National Academies of Sciences, Engineering, and Medicine

12:30-1pm **Break**

1-2:30pm **Session 1B: Intervention Evaluation in Action**

This session will highlight organizations that have had success with aspects of evaluating their interventions, including a walk-through of how they conceptualized their evaluation plans, what they saw as "success," and what progress is being made toward that goal.

- **Michael Leiter**, Honorary Professor at Acadia University and Deakin University
- **Jane Stapleton**, President of Soteria Solutions
- **Melissa Kwon**, Associate Director for Prevention at the PATH to Care Center at the University of California, Berkeley
- **Nicole Merhill** (moderator), Director of the Office for Gender Equity at Harvard University

2:30-3pm **Break**

3-3:45pm   **Session 2: Applying Workshop Concepts and Discussions to Case Studies—Breakout Session**

Workshop participants will reconvene in subgroups to generate a hypothetical concept plan for their case study while having an explicit conversation around addressing limitations, challenges, and barriers.

3:45-4:00pm   **Break**

4:00-4:50pm   **Session 3: Beyond a Workshop: Stakeholder Reflections**

This session will create a space for all workshop participants to collectively reflect on next steps for the workshop's topic. This will be facilitated by a panel of individuals who will share their reflections on the conversations that have taken place, what conclusions they're drawing from the workshop and/or how they might use this information moving forward, and what the future might hold for this topic.

- **Eden King**, Professor of Industrial-Organizational Psychology at Rice University
- **Kiana Swearingen**, Deputy Title IX Coordinator for Education & Prevention at University of Washington
- **Clemencia Cosentino**, Chief Evaluation Officer in the Office of the Director at the National Science Foundation
- **Janine Clayton**, Associate Director for Research on Women's Health at the National Institutes of Health
- **Theodosia Cook**, Chief Diversity Officer at University of Colorado System
- **Larry Martinez** (moderator), Associate Professor and Associate Chair of Psychology at Portland State University

4:50-5pm **Closing Remarks**

National Academies' staff will close the meeting.

- **Arielle Baker**, Program Officer at the National Academies of Sciences, Engineering, and Medicine

Appendix B

Biographical Sketches of Planning Committee Members, Speakers, and Moderators

WORKSHOP PLANNING COMMITTEE

Vicki J. Magley (*Chair*) is a professor in the Department of Psychological Sciences at the University of Connecticut. She has a Ph.D. in social and organizational psychology from the University of Illinois at Urbana-Champaign. The main focus of her research lies within the domain of occupational health psychology and combines both organizational and feminist perspectives in the study of workplace sexual harassment and incivility. She has been studying sexual harassment for 25 years, focusing on its antecedents and consequences for both individuals and organizations, as well as how both individuals and organizations manage sexual harassment. Additionally, she has evaluated sexual harassment awareness training programs and continues her writing on the liability-driven nature of sexual harassment training that is widespread in American companies. She has studied and advocates for civility interventions as an alternative approach to the current typical sexual harassment interventions. Much of her research has resulted from consulting with organizations and federal agencies in understanding their climate of mistreatment and in evaluating interventions designed to alter that climate. She has been a consultant to the World Bank Group, the U.S. Army, the U.S. Air Force, and the National Park Service on sexual harassment climate change efforts and to the U.S. Department of Interior on its self-study of sexual harassment in their bureaus. She has testified to the Department of Defense Judicial Proceedings Panel on the impact of sexual harassment, how feared retaliation inhibits reporting, and the (unfortunate) lack of efficacy of sexual harassment awareness training programs. Most recently, she was one of four research experts on the committee of the National Academies of Sciences, Engineering, and Medicine that produced the 2018 consensus study on sexual harassment in academia. She has been speaking at universities and professional societies about the findings and importance of that report and is on the follow-up advisory committee that is working with 60 universities across the country to implement the recommendations of the report.

NiCole T. Buchanan is a professor in the Department of Psychology at Michigan State University. She has a Ph.D. in clinical-community psychology from the University of Illinois at Urbana-Champaign. Her research focuses on harassment and its impact on organizational climate, employee well-being, and professional development. Specifically, she examines the interplay of race, gender, and victimization and how social identity dimensions affect the nature of harassment; how it is perceived by targets and bystanders; its impact on psychological, occupational, and academic outcomes; and organizational best practices. She also studies barriers to equity and inclusion in higher education: namely, how evaluations of faculty research reflect both discipline-based and identity-based biases that result in the formal and informal epistemic exclusion of marginalized scholars and the scholarship they produce. She is a fellow of the Association for Psychological Science, four divisions of the American Psychological Association (Society of Clinical Psychology, Society for the Psychological Study of Social Issues, Society for the Psychological Study of Ethnic Minority Issues, and Society for the Psychology of Women), and has received national and international awards for her research, teaching, and professional service. She is on the Advisory Group of the Action Collaborative on Preventing Sexual Harassment in Higher Education of the National Academies of Sciences, Engineering, and Medicine. She served on the Research and Content Expert Workgroup for the U.S. Government Accountability Office and the U.S. Congress in conjunction with the National Academies to examine the prevalence and cost of sexual harassment in the U.S. workplace. She has been highlighted in hundreds of media outlets and is a featured speaker in such venues as TEDx and National Public Radio, and she provides bias and diversity-related training and consultation for medical professionals, academic institutions, and other practicing professionals.

Carol W. Greider is a professor of molecular, cellular, and developmental biology at the University of California, Santa Cruz. She has a bachelor's degree from the University of California, Santa Barbara, and a Ph.D. from the University of California, Berkeley. Working together with Dr. Elizabeth Blackburn, she discovered telomerase, an enzyme that maintains telomeres, or chromosome ends, and she shared the Nobel Prize in Physiology or Medicine in 2009 with Dr. Blackburn and Jack Szostak for their work on telomeres and telomerase. Prior to her current position, she held several positions at Cold Spring Harbor Laboratory, first as an independent Cold Spring Harbor fellow, when she cloned and characterized the RNA component of telomerase. She then was appointed as an assistant investigator, followed later by appointment to investigator. She expanded the focus of her telomere research to include the role of telomere length in cellular senescence, cell death, and cancer. Subsequently, she moved her laboratory to the Department of Molecular Biology and Genetics at Johns Hopkins University School of Medicine, where she was appointed as the Daniel Nathans professor and director of the Department of Molecular Biology and Genetics. At Johns Hopkins, her group continued to study the biochemistry of telomerase and determined the secondary structure of the human telomerase RNA. In addition, she characterized the loss of telomere function in mice, which allowed an understanding of short telomere syndromes in humans, such as bone marrow failure, pulmonary fibrosis, and other diseases. She currently directs a group of eight scientists studying both the role of short telomeres in age-related disease and cancer and the regulatory mechanisms that maintain telomere length.

Melissa L. Kwon is the associate director for prevention at the PATH to Care Center at the University of California, Berkeley. She has a bachelor's degree in psychology with minors in ethnic studies and education from the University of California, Berkeley, and a master's degree in education with an emphasis in research methodology and a Ph.D. in education with an emphasis in cultural perspectives and comparative education from the University of California, Santa Barbara. She has dedicated her career to working in and collaborating with higher education institutions, community organizations, and state agencies around preventing domestic, intimate partner, and sexual violence, as well as eliminating oppression. She has worked to prevent domestic and sexual violence at the statewide level through strategic planning, program planning, and collaborative leadership at state agencies, including the Minnesota Office of Higher Education and the Minnesota

Department of Health. She promotes gender equity, racial equity, and education equity through teaching, research, evaluation, and community work.

Larry R. Martinez is an associate professor and associate chair in the Department of Psychology at Portland State University. He earned his B.A., M.S., and Ph.D. degrees at Rice University. His primary area of research is diversity and inclusion in workplace contexts, with a particular emphasis on understanding the experiences of underrepresented, marginalized, and underresearched populations. He also focuses on understanding how employees can be effective allies for one another. This work has received federal funding through a CAREER award from the National Science Foundation. He has published on the topic of sexual and gender harassment, and he participated as an expert panelist at a plenary session at the higher education summit of the National Academies of Sciences, Engineering, and Medicine in Seattle in 2019.

Nicole M. Merhill is director of the Office for Gender Equity and University Title IX Coordinator at Harvard University. She oversees Harvard's overall Title IX system, including direct support and consultation to the university's 50+ Title IX resource coordinators, development and implementation of education programming and prevention efforts, and direct support to community members. She also serves as a representative for the university and a co-lead for the Working Group on Evaluation: Measuring Climate and Gauging Progress on Campus for the National Academies of Sciences, Engineering, and Medicine Action Collaborative, of which Harvard University is a founding member. She also served on the survey design team of the Survey on Sexual Assault and Misconduct of the Association of American Universities, which was administrated at 33 participating institutions and reached 181,752 student respondents. Prior to her position at Harvard, she worked for more than 15 years as a civil rights attorney, with experience in various protected class statuses, including sex, race, color, national origin, disability, and age. She has a bachelor's degree in elementary education and English from Purdue University and a J.D. and a master's degree in education law from the University of New Hampshire School of Law.

WORKSHOP SPEAKERS AND MODERATORS

Kathryn Becker-Blease is an associate professor and director of the School of Psychological Science at Oregon State University. A developmental psychologist, she has researched trauma across the lifespan and effective teaching and learning in higher education. Her most recent work focuses on the intersection of trauma, justice, diversity, equity, and inclusion, and academic success in higher education. She has a Ph.D. in developmental psychology from the University of Oregon.

Jennifer Berdahl is a professor of sociology at the University of British Columbia. Her research tackles issues of gender and power at work, with a focus on sexual harassment and organizational culture. She frequently presents her research to professional, media, and academic audiences and serves as an expert witness in individual and class action discrimination cases in the United States and Canada. She has a Ph.D. in social psychology from the University of Illinois at Urbana-Champaign.

Kelley Bonner is staff at the National Oceanic and Atmospheric Administration, where she works to create and streamline innovative prevention and response services for sexual harassment and sexual assault for all federal employees, contractors, and affiliates at the agency. She has advanced training in workplace resilience and violence prevention. With over 10 years of researching violence trends and working clinically with perpetrators and victims of violence both in the United States and Europe, she believes the key to reducing workplace violence is creating a multidisciplined and comprehensive approach. She is a clinical social worker and holds certifications as a clinical trauma professional in emotional intelligence and mindfulness, and she also holds a master's degree in criminal justice.

Robert Brinkerhoff is an internationally recognized expert in evaluation and learning effectiveness and is also the creator of the success case method, a highly regarded and carefully crafted impact evaluation approach to determining how well a program works. Over his 40-year career, he has provided consultation to dozens of multinational organizations around the globe. He has been a keynote speaker and presenter at hundreds of conferences worldwide and has authored several books, including *The Success Case Method, Telling Training's Story, Courageous Training,* and most recently *Improving Performance Through Learning: A Practical Guide for*

Designing High Performance Learning Journeys. He has a Ph.D. in program evaluation from the University of Virginia and is a retired professor at Western Michigan University, where he coordinated graduate programs in human resource development.

Janine Austin Clayton is associate director for research on women's health and director of the Office of Research on Women's Health (ORWH) at the National Institutes of Health (NIH). She is the architect of the NIH policy requiring scientists to consider sex as a biological variable across the research spectrum. This policy is part of NIH's initiative to enhance research reproducibility through rigor and transparency. As co-chair of the NIH Working Group on Women in Biomedical Careers, with NIH director Francis Collins, she also leads NIH's efforts to advance women in science careers. Prior to joining ORWH, she was the deputy clinical director of the NIH's National Eye Institute. A board-certified ophthalmologist, her research interests include autoimmune ocular diseases and the role of sex and gender in health and disease. She is the author of more than 120 scientific publications, journal articles, and book chapters. She received her undergraduate degree with honors from The Johns Hopkins University and her medical degree from Howard University College of Medicine. She has received numerous awards, including the Senior Achievement Award from the Board of Trustees of the American Academy of Ophthalmology and the European Uveitis Patient Interest Association Clinical Uveitis Research Award. She is also the recipient of a 2010 silver fellow by the Association for Research in Vision and Ophthalmology, the American Medical Women's Association Lila A. Wallis Women's Health Award and the Wenger Award for Excellence in Public Service, and the Bernadine Healy Award for Visionary Leadership in Women's Health. She was an honoree for the *Woman's Day* Red Dress Awards and the American Medical Association's Dr. Nathan Davis Awards for Outstanding Government Service.

Theodosia Cook is chief diversity officer for the four-campus university system of the University of Colorado. In that position she helps develop and implement system-wide policies and initiatives that promote diversity, equity, and inclusion. She coordinates and collaborates with chief diversity officers, faculty, and administrators on each of the campuses. She holds a B.A. in political science from Sewanee: The University of the South and an M.A. in education leadership from Teachers College of Columbia University.

Cindy A. Crusto is a professor of psychology in psychiatry and deputy chair for diversity, equity, and inclusion in the Department of Psychiatry at Yale University School of Medicine; Title IX coordinator at Yale University; and the director of program evaluation and child trauma research at The Consultation Center, a Yale and Connecticut Mental Health Center–affiliated center whose mission is to promote health and wellness, prevent mental health and substance abuse problems, and advance equity and social justice. She is a noted expert in addressing culture, context, and human diversity in clinical work and community-engaged research and program evaluation. She has held leadership roles in the American Evaluation Association, including chairing a task force that developed practice guidelines for addressing culture and context in the profession and in the provision of evaluation services to the public and to evaluation consumers. She has a B.A. in psychology from Vassar College, an M.A. in clinical-community psychology from the University of North Carolina at Charlotte, and a Ph.D. in clinical-community psychology from the University of South Carolina. She completed predoctoral and postdoctoral fellowships in clinical-community psychology in the Department of Psychiatry of the Yale University School of Medicine.

Jennifer M. Gómez is an assistant professor at Wayne State University in the Department of Psychology and Merrill Palmer Skillman Institute for Child & Family Development. She is also a board member and chair of the Research Advisory Committee at the Center for Institutional Courage, a nonprofit institution dedicated to scientific research, wide-reaching education, and data-driven action promoting institutional courage. She is an incoming fellow at the Stanford University Center for Advanced Study in the Behavioral Sciences, as well as the lead co-editor of a special issue of *Journal of Trauma & Dissociation—Discrimination, Violence, & Healing in Marginalized Communities*. Her research on cultural betrayal trauma theory has been recognized by the Ford Foundation and the Michigan Center for Urban African American Aging Research. She has a B.A. in psychology from San Diego State University, and an M.A. in psychology and a Ph.D. in clinical psychology from the University of Oregon.

Lisa M. Hooper has served as an educator, scholar, researcher, mentor, supervisor, and leader for the past 20 years, since receiving her Ph.D. from George Washington University. Early in her career she served as an investigator, project director, and research instructor at Georgetown University

School of Medicine, and later as tenured professor at the University of Alabama and the University of Louisville, directing research focused on the intersection and combined effects of systems (e.g., school, family, neighborhood, community, health care) and race, ethnicity, and culture. Her research constitutes a collaborative, integrative, approach to ecological systems, psychology, education, and whole-person outcomes (e.g., academic, emotional, mental, and physical). The idea of systems and whole-person care has applicability to individuals from cradle to grave and transportability among diverse ecological systems. She has led grants funded by the National Institutes of Health, the National Institute of Mental Health, and, most recently, the United Way to co-design a trauma-informed care program with parents, for parents. She continues to serve as a National Institute on Minority Health and Health Disparities scholar. Recently, she collaborated with the Office of Minority Health at the U.S. Department of Health and Human Services to create an educational initiative focused on cultural and linguistic competence.

Bianca Kaushal-Carter is the manager for prevention education in the Institute Discrimination and Harassment Response Office at the Massachusetts Institute of Technology (MIT). She earned her M.S.W. from the Brown School of Social Work at Washington University in St. Louis where her research and practical training focused on systems approaches to violence and injury prevention. At MIT, she has worked with campus partners to develop the promoting inclusive environments workshops and deliver training, both in-person and online, to more than 10,000 students, faculty, and staff members annually. In addition to these initiatives, she also serves as a staff representative on the MIT Committee for Sexual Misconduct Prevention and Response and works with the MIT implementation team to advance recommendations created in response to the *Sexual Harassment of Women: Climate, Culture, and Consequences in Academic Sciences, Engineering, and Medicine* (National Academies of Sciences, Engineering, and Medicine, 2018).

Eden King is Lynette S. Autrey professor of industrial-organizational psychology at Rice University. She is pursuing a program of research that aims to make work better for everyone. This research—which has yielded over 100 scholarly products and has been featured in outlets such as *The New York Times*, Good Morning America, and *Harvard Business Review*—

addresses three primary themes: (1) current manifestations of discrimination and barriers to work–life balance in organizations, (2) consequences of such challenges for its targets and their workplaces, and (3) individual and organizational strategies for reducing discrimination and increasing support for families. In addition to her scholarship, she has partnered with organizations to improve diversity climate, increase fairness in selection systems, and to design and implement diversity training programs. She is currently co-editor of the *Journal of Business and Psychology* and immediate past president of the Society for Industrial and Organizational Psychology.

Kurt Kraiger is a professor and chair of the Department of Management at the Fogelman College of Business and Economics at the University of Memphis. He received a B.A. in psychology from the University of Cincinnati and a Ph.D. in industrial-organizational psychology from The Ohio State University. He has published extensively and presented workshops in areas related to workplace training generally and training evaluation specifically.

Michael P. Leiter is honorary professor of organizational psychology at Deakin University in Melbourne, Australia. Previously, he held the Canada research chair in occupational health at Acadia University. He has published widely on job burnout, work engagement, and workplace civility, with over 98,000 citations on Google Scholar. His recent initiatives have focused on improving the quality of work life through enhancing civility and respect among colleagues as a strategy for reducing burnout. He currently writes and consults through Michael Leiter & Associates in Nova Scotia, Canada. He has a B.A. from Duke University, an M.A. from Vanderbilt University, and a Ph.D. from the University of Oregon.

Elissa L. Perry is a professor in the Social-Organizational Psychology Program at Teachers College at Columbia University. Her research focuses on the role of demographic characteristics (age, gender, disability, race) in human resource judgments and organizational behavior. She is also interested in strategies (training, organizational climate change, inclusive leadership) for managing diversity and sexual harassment. Her most recent and active scholarship is in the areas of inclusive leadership and the impact of organizational racial and gender diversity on organizational outcomes. She has a Ph.D. in organizational psychology and theory from Carnegie Mellon University.

Alec M. Smidt is a postdoctoral associate in the Department of Psychiatry at the Yale University School of Medicine. He received his Ph.D. from the University of Oregon and completed his residency at Yale. His research focuses on the effects of interpersonal and institutional betrayal on those who have experienced these harms. Of particular interest to him is how institutional betrayal may exacerbate the effects of interpersonal trauma and how certain groups (e.g., women and sexual and gender minorities) may be more at risk than others for experiencing both interpersonal and institutional betrayal in a variety of institutional contexts.

Raechel Soicher is an instructor in the School of Psychological Science at Oregon State University. She recently received a Ph.D. after defending her dissertation, "Implementation Science in Teaching and Learning in Higher Education: A Mixed-Methods Study of the Utility Value Intervention in General Psychology Courses." As an applied psychologist, her interdisciplinary research focuses on applying an implementation science lens to the translation of cognitive psychology to promote evidence-based teaching in higher education.

Jane G. Stapleton is executive director of practice at the University of New Hampshire Prevention Innovations Research Center and the co-founder and president of Soteria Solutions, which provides research-based, strategic, and thoughtful approaches to prevent sexual harassment and interpersonal harassment. At Soteria she directs curriculum development, technical assistance, and training that creates lasting change by building safe and respectful workplaces and learning environments. Her work focuses on translating research to practice and using data to ensure that prevention strategies meet target populations' specific needs and characteristics. She is co-author of Bringing in the Bystander*, a scientifically evaluated and proven-effective prevention program for workplaces, colleges and universities, and high schools. She also co-developed the Know Your Power* Bystander Intervention Social Marketing Campaign. She is the project director for Soteria's collaboration with the National Oceanic and Atmospheric Administration (NOAA) to assess climate and build NOAA-specific prevention strategies.

Karen R. Stubaus serves as vice president for academic affairs at Rutgers, The State University of New Jersey, responsible for a broad array of academic and strategic matters across the University's three geographical locations in New Brunswick, Newark, and Camden. She has been a leader in increasing the diversity of the faculty and in promoting women's leadership at all levels of the institution. She is centrally responsible for faculty labor relations, for the university's initiative on sexual harassment prevention and culture change, and for increasing professional development opportunities for contingent faculty. She also teaches the Department of American Studies of the School of Arts and Sciences on both the undergraduate and graduate levels. She is currently engaged in a study of the intersection of graduate student activism, sexual harassment, and collective bargaining. She has a Ph.D. in 17th-century American history.

Kiana Swearingen is deputy Title IX coordinator for education and prevention at the University of Washington. She has worked in the anti-violence field for 15 years and in higher education for 6 years. Her work is centered on intersectional and trauma-informed prevention and response to sexual harassment, relationship violence, stalking, sexual exploitation, and sexual assault. She believes that successful approaches to preventing sex- and gender-based violence must engage individuals as agents of change, give communities practical strategies to build respectful and inclusive environments, and analyze the structures and policies that perpetuate and exacerbate harm. She has trained regionally and nationally on theory-driven community-centered best practices for developing, conducting, and evaluating violence prevention efforts and building innovative systems-level responses for survivors.

Nelia Viveiros is associate vice chancellor for academic operations at the University of Colorado, Anschutz Medical Campus. A first-generation, Latinx leader whose career in higher education has spanned over 20 years, she oversees a range of matters relating to Title IX, Title VII, equity and compliance, conflict de-escalation, training, and prevention programming across the University of Colorado, Anschutz Medical Campuses.

Appendix C

Workshop Participants

Name	Title	Organization
Shafiqa Ahmadi	Professor	University of Southern California
Elizabeth Armstrong	Sherry B. Ortner Collegiate Professor of Sociology	University of Michigan
Victoria Banyard	Professor and Associate Dean of Faculty Development	Rutgers University School of Social Work
Kathryn Becker-Blease	Associate Professor and Director, School of Psychological Science	Oregon State University
Jennifer Berdahl	Professor	University of British Columbia
Mindy Bergman	Professor	Texas A&M University
Alan Berkowitz	Independent Consultant	Independent Consultant
Meg Bond	Professor of Psychology, Director, Center for Women & Work	University of Massachusetts Lowell

Name	Title	Organization
Kelley Bonner	Director, Workplace Violence Prevention & Response Office	National Oceanic and Atmospheric Administration
Robert Brinkerhoff	Professor Emeritus; Director of Research and Evaluation	Western Michigan University; Brinkerhoff Evaluation Institute & Promote International
NiCole Buchanan	Professor	Michigan State University
Lisa Bundesen	Senior Policy Advisor	National Institutes of Health
Rebecca Campbell	Professor	Michigan State University
Kathryn Clancy	Associate Professor of Anthropology	University of Illinois at Urbana-Champaign
Janine Clayton	Associate Director for Research on Women's Health	National Institutes of Health
Megan Clifford	Associate Laboratory Director	Argonne National Laboratory
Theodosia Cook	Chief Diversity Officer	University of Colorado System
Clemencia Cosentino	Chief Evaluation Officer, Office of the Director	U.S. National Science Foundation
Cindy Crusto	Associate Professor	Yale University
Katie Edwards	Associate Professor	Prevention Innovations Research Center, University of New Hampshire
Armando Estrada	Associate Professor of Policy, Organizational & Leadership Studies	Temple University
Amanda Field	Health Science Policy Analyst	National Institutes of Health

APPENDIX C   *81*

Name	Title	Organization
Kathy Friedman	Associate Professor and Vice Chair, Biological Sciences	Vanderbilt University
Jessica Gallus	Senior Advisor	U.S. Department of the Navy
Christine Gidycz	Professor Emerita	Ohio University
Jennifer M. Gómez	Assistant Professor	Wayne State University
Carol Greider	Distinguished Professor	University of California, Santa Cruz
Sarah Harebo	UW System Title IX and Clery Administrator	University of Wisconsin System
Lisa Hooper	Professor and Richard O. Jacobson Endowed Chair for Research	University of Northern Iowa
Emily Huang	Associate Professor	Oregon Health and Science University
Liz Hutchison	Associate Vice President for Equity and Inclusion	University of New Mexico
Colleen Johnston	Title IX Coordinator	Northwestern University
Lisa Kath	Associate Professor	San Diego State University
Bianca Kaushal-Carter	Manager of Prevention, Education, and Outreach	Massachusetts Institute of Technology
Taek K. Kim	Nuclear Systems Analysis Department Manager	Argonne National Laboratory
Eden King	Professor	Rice University
Kurt Kraiger	Professor of Management	University of Memphis

Name	Title	Organization
Melissa Kwon	Associate Director for Prevention, PATH to Care Center	University of California, Berkeley
Diana Lautenberger	Director of Faculty and Staff Research and Gender Equity Lab	Association of American Medical Colleges
Michael Leiter	Honorary Professor	Acadia University, Canada, and Deakin University, Australia
Vicki Magley	Professor of Psychological Sciences	University of Connecticut
Larry Martinez	Associate Professor, Associate Chair, Psychology	Portland State University
Brian C. Martinson	Senior Research Investigator	HealthPartners Institute
Sara McClelland	Associate Professor, Psychology and Women's & Gender Studies	University of Michigan
Sarah McMahon	Associate Professor	Rutgers University School of Social Work
Nicole Merhill	Director of the Office for Gender Equity	Harvard University
Lynn Morin	Health Scientist Administrator	National Institutes of Health
Kaylin Padovano	Staff and Faculty Training Coordinator	Rutgers University
Andrea Page-McCaw	Professor of Cell and Developmental Biology	Vanderbilt University School of Medicine
Elissa Perry	Professor of Psychology and Education	Teachers College, Columbia University
Sharyn Potter	Professor/Executive Director of Research	University of New Hampshire Prevention Innovations Research Center

Name	Title	Organization
Sarah Rankin	Director and Title IX Coordinator	Massachusetts Institute of Technology
Paige Sechrest	Prevention, Education, and Communications Manager	University of Washington
Alex Smidt	Postdoctoral Associate	Yale University School of Medicine
Raechel Soicher	Instructor, School of Psychological Science	Oregon State University
Jane Stapleton	President	Soteria Solutions
Karen Stubaus	Vice President for Academic Affairs	Rutgers University
Kiana Swearingen	Deputy Title IX Coordinator for Education & Prevention	University of Washington
Elizabeth Umphress	Professor	University of Washington
Nelia Viveiros	Associate Vice Chancellor for Academic Operations	University of Colorado, Anschutz Medical Campus
Quinn Williams	General Counsel	University of Wisconsin System
Alexandra Zelin	Assistant Professor	University of Tennessee at Chattanooga

Appendix D

Workshop Case Studies

This appendix comprises six "case studies" that were sent to participants prior to the workshop. Each case study contains a 2–5-page description of an actual program, policy, or practice currently being carried out at an institution. This information comes from members of the Action Collaborative on Preventing Sexual Harassment in Higher Education (see Chapter 1); they submit these "Descriptions of Work" annually, which are then made public by the National Academies of Sciences, Engineering, and Medicine.[1] For the purposes of this workshop and proceedings, these descriptions have been lightly edited for clarity.

The Action Collaborative created a rubric to help organizations identify areas of work that are consistent with the findings and recommendations of *Sexual Harassment of Women* (National Academies of Sciences, Engineering, and Medicine, 2018). The rubric comprises 28 activities in four categories: prevention, response, remediation, and evaluation.

To facilitate the sharing of actions taken, potential innovative ideas, and research on the effectiveness of actions, in the summer of 2020, Action Collaborative members were asked to provide at least one and up to five descriptions of their most significant, innovative efforts—either in the planning stages or being implemented—that are consistent with the report's

---
[1] Available: www.nationalacademies.org/sexual-harassment-collaborative-repository.

findings and recommendations and are new for either the organization or higher education overall.

At the workshop, these case studies were incorporated into presentations and used by breakout groups so that participants could consider whether and how they could be evaluated. Each participant was asked to read one case study, and they were welcomed to read more than one.

The purpose of using these case studies was not to judge them on their merit or likelihood for success; rather, the purpose was to help participants consider how actual prevention efforts might be evaluated for effectiveness.

Case Study A	**A Guide on Best Practices for Graduate Students Impacted by Sexual Violence and Sexual Harassment (SVSH)**—University of California, Santa Cruz
Case Study B	**Addressing Harassment in Employment Practices**—University of Washington
Case Study C	**Altering Departmental Admissions Policies to Diffuse Dependent Relationships Between Graduate Students and Their Advisors**—Vanderbilt University
Case Study D	**Core Values Shout-Outs**—Argonne National Laboratory
Case Study E	**Lab-Based Inclusive Culture Workshops**—Massachusetts Institute of Technology
Case Study F	**Significantly Increasing Faculty and Staff Education and Skill Development**—Rutgers University

CASE STUDY A

UNIVERSITY OF CALIFORNIA, SANTA CRUZ

A GUIDE ON BEST PRACTICES FOR GRADUATE STUDENTS IMPACTED BY SEXUAL VIOLENCE AND SEXUAL HARASSMENT (SVSH)

This Action Applies to Three Rubric Items in the Remediation Category: preventing retaliation, reintegration strategies and programs, and other efforts to remediate the harm of sexual harassment and/or support those that experience sexual harassment.

Description of Work

Several high-profile instances of sexual violence/sexual harassment (SVSH) at our institution have involved interactions between faculty and graduate students. More broadly, as recognized in the report of the National Academies of Sciences, Engineering, and Medicine (NASEM, 2018) and other documents, several characteristics of these relationships (male-dominated in some disciplines; hierarchical and dependent relationships between faculty members and graduate students; isolating environments) create higher levels of risk for sexual harassment. And because of the dependence of graduate students on faculty mentors for different types of support, guidance, and promotion in the discipline, the disruption for a student associated with both the reporting of SVSH and the investigation and outcome of a report can be great.

The Beyond Compliance initiative at University of California, Santa Cruz, which was formed in 2016, seeks to define concrete mechanisms that administrators and faculty can adopt, above and beyond what is required by law or policy, to address SVSH on campus. The committee is co-chaired by a senior faculty member and a senior administrator, and it includes faculty members, graduate students, and key staff who are working to address SVSH and its consequences. The committee recognized the cascading impacts of SVSH on graduate students and decided to assemble a quick guide on best practices to remediate these impacts.

The guide considers a range of potential negative consequences for a graduate student who has experienced and reported SVSH. These include loss of the individual's advisor, classroom professor, committee member, or letter writer; financial impacts due to loss of teaching-assistant or research-assistant support, as well as health insurance if the student takes a leave, slowed academic progress, challenges meeting residency requirements if the student moves away from campus to access support, and scenarios around journal submissions and grant proposals. For each potential impact, the guide identifies a point person or persons and any units with ultimate responsibility for ensuring that the issue is addressed, from the chair or director of graduate studies in the department, to the divisional dean, to the dean of graduate studies, to the campus provost/vice chancellor. The guide also identifies the office to which graduate students should report different types of conduct violations. Finally, for each type of impact, the guide suggests the types of remediation the point person(s)/unit should consider and who they might consult in moving forward with remediation. Below is an example of one of the entries on the spreadsheet.

Issue	Point People (person/unit responsible for ensuring situation is addressed)	Others Who May Need to be Involved	Recommended Solution/ Menu of Solutions
Student drops a class (e.g., due to strain of P&T hearing, Title IX investigation) which threatens the student's continued financial aid/scholarship status during the subsequent term.	Graduate Division (via Vice Provost and Dean of Graduate Studies)	Request could come in via multiple paths (Dean, chair, Title IX, etc.).	Graduate Division is likely in the best position to coordinate such exceptions/appeals with the Financial Aid office. If it becomes a situation of "covering" the cost of a scholarship etc. that may be best handled by Campus Provost/Executive Vice Chancellor.

Next steps: The guide was finalized in the middle of the 2019–2020 academic year and sent to the Division of Graduate Studies for comment. Due to a confluence of high-profile events (a wildcat graduate student strike beginning in November, the global coronavirus pandemic, campus upset and action around the killing of George Floyd), the Graduate Division has not yet responded to Beyond Compliance. We will request comments by early fall 2020. Following this review, we will send the guide to the Academic Senate for its input (certainly to the Committee on Affirmative Action and Diversity and the Graduate Council, and perhaps the Committee on Research). When Senate review is complete, the document will be finalized and shared with all units that bear responsibility for remediating impacts on graduate students, as well as with the Graduate Student Union.

Website for further information (if applicable):
Point of Contact Name: Paul Koch
Email Address for Point of Contact: plkoch@ucsc.edu

> # CASE STUDY B
>
> # UNIVERSITY OF WASHINGTON
>
> ## ADDESSING HARASSMENT IN EMPLOYMENT PRACTICES

This Action Applies to One Rubric Item in the Response Category: improved communication and increased transparency.

Description of Work

Washington State Engrossed House Bill 2327 went into effect June 11, 2020. University of Washington (UW) staff, faculty, and other academic personnel, and community members provided input on the bill and testified in front of legislative committees in support of the bill. The law addresses many of the priorities outlined in the 2018 report of the National Academies of Sciences, Engineering, and Medicine. For example, the bill requires institutions of higher education in the state of Washington to share findings of harassment during reference checks, to ban the use of nondisclosure agreements, and to share publicly the results of any climate surveys. The state law, as described below, is specifically in response to harassment of students by employees; however, UW is expanding our procedures and compliance to include harassment of students *and* employees. Further information about the law follows.

Investigations and records: While UW had already made a commitment to completing investigations, the law now requires that unless a victim requests otherwise, a postsecondary institution will complete investigations of alleged sexual misconduct regardless of whether the employee voluntarily or involuntarily leaves employment with the institution. Written findings of completed investigations are required, and substantiated findings against an employee will be included in the employee's personnel file. If, at the end of an investigation, allegations have not been substantiated, information about the allegations may be expunged.

Declaration of applicants: Prior to an offer of employment, applicants are required to sign a statement declaring whether they have ever been found responsible for any sexual misconduct at their current or a prior place of employment and, if so, an explanation of what happened. The declaration also requires the applicant to indicate if an investigation is currently under way and to authorize current and past employers to disclose substantiated misconduct or current investigations. If applicants do not sign this statement, institutions cannot hire them.

Required reference checks: Prior to hiring an applicant, colleges and universities must request from any postsecondary institutions at which the applicant currently works or has worked any records about any substantiated findings of sexual misconduct. If information obtained regarding an applicant indicates any issue of concern, UW will take appropriate next steps.

The law also requires all postsecondary institutions to respond to reference checks from other postsecondary educational institutions by disclosing any substantiated findings of sexual misconduct and to provide information (documents, records, etc.) about any substantiated findings of sexual misconduct to any employer inquiring about one of their applicants regardless of whether such information is specifically requested. In order to comply with this section of the law, each institution must establish procedures to provide information about substantiated findings of sexual misconduct when another institution requests any such findings and ensure that those procedures will not disclose identifying information about a complainant or witnesses.

Ban nondisclosure agreements: The law states that settlement agreements may not include provisions prohibiting employees, institutions, survivors, or others from disclosing information that an employee is or was the subject of an investigation about sexual misconduct or of an investigation that yielded findings of sexual misconduct. Settlement agreements may include provisions allowing nondisclosure of identifying information about complainants or witnesses. Identifying information of complainant/witnesses also is not subject to disclosure through public records requests.

Share results of climate assessments: When a climate assessment is conducted, postsecondary institutions need to share with the governor and appropriate legislative committees summaries of any climate assessment designed to gauge prevalence of sexual misconduct. In addition, institutions must include a description of efforts to reach out to and obtain information from traditionally marginalized students or those who disproportionately experience impacts of systemic oppression based on protected categories (e.g., race, ethnicity, nationality, sexual orientation, gender identity, gender expression, and disability), along with information about how the results of the assessment were used to design or improve policies, programs, and resources for the campus community.

This Washington state law went into effect in June 2020 with an October 1, 2020, compliance date for applicant declaration, a July 1, 2021, compliance date for the reference check requirement, and a December 2023 date for the sharing of any climate assessment information. UW is currently working to establish procedures within Human Resources and the Office of Academic Personnel that includes, among other things, centralizing records that contain findings and ensuring that they can be shared in compliance with state law.

Website for further information (if applicable):
http://lawfilesext.leg.wa.gov/biennium/2019-20/Pdf/Bills/Session%20Laws/House/2327-S.SL.pdf?q=20200416122655
Point of Contact Name: Valery Richardson
Email Address for Point of Contact: valeryr@uw.edu

CASE STUDY C

VANDERBILT UNIVERSITY

ALTERING DEPARTMENT ADMISSIONS POLICIES TO DIFFUSE DEPENDENT RELATIONSHIPS BETWEEN GRADUATE STUDENTS AND THEIR ADVISORS

This Action Applies to One Rubric Item in the Remediation Category: Reducing power differentials

Description of Work

The Problem: Admission to Graduate School to Work with a Specific Advisor

Graduate programs in biological/biomedical sciences may use any of three distinct types of admission processes. The most common is for several programs to band together and offer admission through an umbrella program, in which accepted students share a common first-year curriculum, meet several potential advisors from many graduate programs, and then join an advisor's lab within a graduate program. Vanderbilt has two such umbrella programs, the Interdisciplinary Graduate Program (IGP) and the Quantitative and Chemical Biology Program (QBP) and all of Vanderbilt's biomedical graduate programs participate.

In a second route, a graduate program may accept students into its specific program, in which students similarly can meet with several potential advisors and choose one. Vanderbilt's Biological Sciences Department and Graduate Program in Neuroscience utilize this model, as well as the umbrella programs.

The third process is for a student to apply to a graduate program to work with one specific advisor, chosen in advance of applying. This is a direct-admit process. Sometimes the student and advisor have worked together previously, and sometimes they are unknown to each other. Direct-admit students may have less information for choosing an advisor, and they have less power to change labs if problems arise. Eight of the biomedical graduate programs at Vanderbilt offer students a direct-admit path to entry, and one of these is the Department of Cell and Developmental Biology (CDB).

In 2019, CDB appointed a new director of graduate studies (DGS), who was also independently appointed by the provost to serve as a co-delegate to the Action Collaborative. The new DGS observed that direct-admit students seemed to have a more difficult path in graduate school than other students, with more advisor-student conflict and increased rates of leaving the Ph.D. program before graduation.

Structural differences in the direct-admit experience in CDB before the policy changes:

- Admission: Umbrella students are interviewed by four faculty members and reviewed by an admissions committee before being accepted. In contrast, direct-admit students were vetted by the prospective advisor without a standard framework.
- First-year funding: Students who enter through umbrella programs have their first year paid by university funds; they switch to advisor-based funding after they choose a lab. Direct-admit students were funded by their advisor from the first day in the program.
- Rotations: Students who enter through umbrella programs are required to do four rotations, in which they "try out" working in four different labs for a period of about 2 months in each lab, with the goal of identifying a thesis laboratory. These students experience a variety of mentoring styles, get a sense of cultural norms, and develop a professional network. In contrast, direct-admit students work with the same advisor's lab from the beginning.
- Social experience: Students who enter through umbrella programs have a class identity and usually form strong bonds with classmates, forged in part in mentoring groups, which are a required part of their programs. In contrast, direct-admit students are more isolated and often know only people in their own lab. They are not required to participate in the mentoring groups because of the attention given in these groups to identifying a thesis laboratory.
- Switching labs: If a student requests, the umbrella program will pay for a student to undertake more rotations to find a new lab if problems arise. Further, students who enter through umbrella programs can enlist the help of the umbrella program faculty advisors and former rotation advisors in identifying a new lab. In contrast,

the direct-admit students have few human resources, no access to funded rotations, and do not carry the official stamp of acceptance of the umbrella program.

Report of the National Academies of Sciences, Engineering, and Medicine on sexual harassment of women—Analysis and relevant recommendations:

The report's Recommendation 5 is *to diffuse the hierarchical and dependent relationship between trainees and faculty.* Direct-admit students are particularly dependent on their advisor because of the complete dependence on the advisor's funding, their reduced professional network, and their decreased potential to change labs. Specifically, the report recommends mentoring networks and departmental funding.

The process of shaping a new policy:

1. Information was collected (May 2020) on all CDB direct-admit students for the last 11 years.
    a. Direct-admit students accounted for approximately 10% of doctoral students during this time.
    b. 40% of direct-admit students were foreign nationals. Because of visa stipulations that they must leave the United States if they leave graduate school, foreign nationals would be a uniquely vulnerable group. Further, they have fewer social supports in this country.
    c. 30% of direct-admit students had worked with their advisor before admission to graduate school as an undergraduate, a summer intern, or an employee. These students were very successful and had few problems.
    d. Direct-admit students were more likely than umbrella students to leave graduate school without a Ph.D. or to have significant academic problems. (We are not reporting these percentages because of the small number of students involved.)
    e. Anecdotally, it was frequently observed that direct-admit students did not know about the existence of the umbrella programs; or they did not realize that by choosing a direct-admit pathway their options were limited compared with their peers.
2. The other 10 biomedical graduate programs at Vanderbilt were queried by email about whether they had direct-admit programs

and how they managed them: "*Will your program accept students to work directly with one faculty member who is entirely responsible for their finances?*" Eight programs used the direct-admit path at varying frequencies and with various admissions requirements. The DGS discussed with the leaders of these programs their experiences with direct-admit programs (June 2020). Several focused on the success of students who had worked in the lab prior to matriculation.

3. The DGS and the CDB department chair met to discuss the problem and outline potential solutions, including departmental funding.
4. The DGS convened a meeting of the CDB Graduate Education Committee, composed of six CDB faculty (at that time. three assistant professors and three full professors) to brainstorm and evaluate solutions.
5. The Graduate Education Committee negotiated and compromised over email for several weeks, arriving at a new set of policies with unanimous support.
6. The committee recommended these policies to the full faculty, who voted overwhelming in favor (August 2020). The full policy can be found here: https://medschool.vanderbilt.edu/cdb/admission-to-cdb-graduate-program.

Why the direct-admit path was reformed rather than abolished:

Although the DGS originally suggested eliminating the direct-admit program entirely, the Graduate Education Committee did not support this approach for the following reasons.

1. Many direct-admit students with prior experience in the lab were spectacularly successful and clearly understood the environment they were joining.
2. The high proportion of foreign nationals was viewed as a benefit. Because of issues with funding international students, the IGP and QCB umbrella programs admit a limited number of foreign nationals in each class. Thus, abolishing the direct-admit path in CDB would reduce access for international students to an Ameri-

can education, reduce access of the labs to highly talented students, and reduce CDB's multicultural diversity.
3. Some faculty in CDB rely on direct-admit graduate students for their labs. Abolishing this route of admission would disproportionately affect them and likely interfere with their research programs.

Key elements of the new policy:
Policies were put in place to promote these goals:

1. To preserve the student's choices. Because a student spends about 6 years in a lab with one advisor, they should know what they are signing up for.
    a. Students must apply to a Vanderbilt umbrella program in order to be considered for direct-admission to a lab at CDB. Only application, not admission, is required. By applying to the umbrella programs, students will become aware of their benefits and may choose that route of entry if available. Umbrella programs are the preferred route for graduate school, as they offer students the most choices and resources.
    b. Students expressing interest in the direct-admit program will be advised about the differences between it and the umbrella programs by the DGS.
    c. Students are expected to work in an advisor's lab before applying to that lab through the direct-admit program. If they have not, they must explain why this is not possible in their application materials.
    d. Mentors must send their Mentoring Compact (a document outlining the responsibilities of the mentor and student) to the student before an offer of admission can be extended.
    e. If a student wants to change labs, and if this decision is supported by the pre-candidacy committee (see below), the department will fund the student for 12 weeks, which is sufficient time to do two rotations.
2. To increase the professional and social network of the student.
    a. A pre-candidacy committee will work with the student from the first week of the first year until a thesis committee is formed. The committee is charged with overseeing the student's intellectual growth, class performance, lab productivity, and social adjustment to graduate school.

  b. A student-advisory group will be assigned to the student on arrival to campus, composed of three students with various overlapping interests, to provide a social network for the student.
  c. Direct-admit students will participate in mentoring groups with the umbrella students.
3. To maintain admissions standards.
  a. The application will be reviewed by the CDB Graduate Education Committee rather than just by the prospective mentor. Because candidates will have applied to an umbrella program, feedback from that admissions program can be taken into consideration by the committee. The ultimate admissions decision will be made by the committee rather than just by the DGS. This will make it easier for a student to change labs if needed.

Future evaluation metrics:

 There are three metrics by which to measure success of the policy.

 The first and most important is the experience of the direct-admit graduate students for the next 6 years, which is approximately one graduate-student generation. Because the numbers are expected to be small, this measure will be qualitative, noting their connectedness, levels of conflict with their advisor, academic success and professional productivity, and retention to completion of a Ph.D.

 The second is to measure how many direct-admit graduate students have previously worked with their advisor prior to arriving. An increase in this percentage will be considered a success of the policies.

 The third measure is the overall number of direct-admit graduate students. A decrease in their number will be considered a success of the policies.

Website for further information (if applicable): https://medschool.vanderbilt.edu/cdb/admission-to-cdb-graduate-program
Point of Contact Name: Andrea Page-McCaw
Email Address for Point of Contact: andrea.page-mccaw@vanderbilt.edu

CASE STUDY D

ARGONNE NATIONAL LABORATORY

CORE VALUES SHOUT-OUTS

This Action Applies to One Rubric Item in the Prevention Category: civility or respect promotion programs.

Description of Work

In July 2018 Argonne National Laboratory established a set of core values as the foundation of the laboratory's efforts to help create and sustain a safe, welcoming, diverse, and inclusive environment that enables all members of the Argonne community to perform their best work.

To maintain awareness of Argonne's Core Values of Impact, Safety, Respect, Integrity, and Teamwork, as well as to highlight positive behaviors related to each value and promote action, Argonne created an employee engagement program called Core Values Shout-Outs. The program encourages employees to recognize colleagues for demonstrating the core values through their behaviors. This aligns with recommendations in the 2018 report of the National Academies of Sciences, Engineering, and Medicine that anti-harassment efforts be combined with civility or respect promotion programs as a mechanism for highlighting behaviors that faculty, staff, and students should engage in rather than focusing only on negative behaviors.

The Core Values Shout-Outs program kicked off on July 31, 2019, in conjunction with the unveiling of a set of most valued behaviors for each core value. The Shout-Outs program was created as a simple way for employees to learn about and recognize valuable behaviors and to reinforce the shared accountability of each member of the Argonne community for creating a safe, welcoming, diverse, and inclusive work environment.

To give a colleague a shout-out, employees were asked to submit a simple online form with the recipient's name, the core value they exemplified, and a one- or two-sentence description of the positive behavior they modeled. Awardees and their supervisors automatically received an email notification. The recipients then received a button for that value. Information from the shout-outs submissions was collected in a database that was shared with the laboratory's Core Values Working Group and with laboratory leadership.

Thanks to the simplicity of giving a shout-out and the immediacy of the recognition, the Shout-Outs program was embraced wholeheartedly by employees at all levels of the laboratory. Within weeks, it became common to see employees proudly displaying the buttons they had earned on their lanyards. The campaign increased the visibility of Argonne's core values and positive behaviors not only through buttons on lanyards, but also through an internal communications promotion of the program, which included digital posters, stories in the daily email employee newsletter, and information on the employee intranet. Volunteer core values ambassadors in every division at the laboratory not only helped to facilitate the program, but also provided increased visibility and promotion.

Originally set to run from July 31 through December 31, 2019, the program was extended to March 5, 2020, due to its popularity and the value laboratory leadership saw in it as a tool for recognition and encouragement of positive behavior. At a lab-wide, all hands meeting on November 14, 2019, Argonne director Paul Kearns challenged employees to achieve 3,000 shout-outs by the end of the program.

The laboratory community rose to the challenge, and the program ended March 5, 2020, with more than 3,400 shout-outs given. An analysis of the data collected through the program provided further insights that were shared at a lab wide all-hands meeting on May 28, 2020, through a light-hearted video featuring employees from across the laboratory. This included learning that:

- Every division across the lab gave and received shout-outs.
- The core value recognized the most was teamwork.
- More than 85% of shout-outs received were given peer to peer rather than by supervisor to staff.

Data collected from the Shout-Outs program also was used to create case studies for discussion during training offered to all employees on learning how to identify and address situations and behaviors that are not consistent with the lab's core values. This training will be offered through August 2020.

Further evaluation of the core values and the Shout-Outs program was sought in a pulse employee climate survey conducted over 3 weeks in June of 2020. Specifically, respondents were asked about activities that have positively influenced their thinking or behavior including the laboratory's focus on Core Values efforts and programs such as the Shout-Outs. Survey results are expected in late summer 2020.

As a majority of the laboratory's employees moved to teleworking in March 2020 as a result of the COVID-19 pandemic, laboratory leadership identified a need to keep the core values visible to employees working remotely. Plans were developed to bring the Core Values Shout-Outs back but in a completely digital format.

Shout-Outs 2.0, the digital version, was introduced at a lab-wide all-hands meeting on May 28, 2020, to help emphasize the importance of employees staying connected while working apart. Through the laboratory human resources application Workday, employees still submit shout-outs. Instead of a physical button, a digital button appears in the recipient's Workday account. Both recipients and their supervisors receive notification of the shout-out.

Shortly after the Shout-Outs 2.0 roll out, the information from the more than 3,400 shout-outs in the previous campaign was merged with the new, digital version. Now, employees can view all the shout-outs they've earned in one place in Workday. Work is currently under way to create further online visibility for each employee's shout-outs. In fall of 2020, shout-outs will be displayed on employees' individual intranet homepages, in their online profiles, and in the laboratory's online directory.

What began as a means to create awareness of the laboratory's core values and enlist employee participation in putting the values into action has evolved into a continued form of recognition and a constant, visible reminder of what Argonne values as a community. This further solidifies the core values as the foundation of the laboratory's efforts to achieve its goals of expanding Argonne's leadership in science and technology, achieving operational excellence, and building and sustaining a world-class community of talent.

Website for further information (if applicable): https://www.anl.gov/our-core-values
Point of Contact Name: Megan Clifford
Email Address for Point of Contact: mclifford@anl.gov

CASE STUDY E

MASSACHUSETTS INSTITUTE OF TECHNOLOGY (MIT)

LAB-BASED INCLUSIVE CULTURE WORKSHOPS

This Action Applies to Five Rubric Items, Four in the Prevention Category and One in the Response Category: civility or respect promotion programs, leadership education and skill development, bystander intervention programs, audience-specific anti-sexual harassment education, and addressing gender harassment and other bad behaviors.

Description of Work

1. **The purpose and goals of what you did or what you are doing, and how you did it**

   Lab-based workshops were developed as an initiative to reach graduate students and postdocs and train them about issues and resources related to gender bias and sexual harassment. The lab workshops were based on prevention research that shows that conducting workshops with intact groups is an effective approach for shifting culture and building skills. The lab serves as an important unit of community within the Massachusetts Institute of Technology (MIT) environment, and the lab workshops provide an opportunity for MIT resources to connect with the lab and help reestablish norms. It's important to note that the offices that have previously led this initiative (the Institute Discrimination and Harassment Response Office [IDHR] and Violence Prevention and Response [VPR]) approached this work at the department-level instead of by request from individual labs to amplify the impact of a department initiative. These offices would partner with the students and department leadership, conduct a faculty demo during a regular faculty meeting, and then the department chair would put it to a vote (once the presenters left) so that this could be a *faculty-driven* initiative with the appropriate buy-in and engagement. Additionally, most department heads would require the workshop to be completed in a certain time frame by all faculty in the department.

2. **How it is consistent with the findings and recommendations of the 2018 Report of the National Academies of Sciences, Engineering, and Medicine (as outlined in the rubric)**

   MIT partners with departments to develop a tailored, 2-hour in-person workshop that is delivered by trained facilitators to each lab cohort, "Promoting a Professional and Inclusive Lab Culture." Attendance at each workshop includes all students, postdocs, and the principal investigator (PI). Typically, each PI in the department is required to host a workshop at some point during a designated semester. The workshop content is created using focus group feedback from students and staff about climate issues within the department and leverages any school/department level climate data, and national field data (if available). Content includes:

   - The impact of unintentional harms/micro-aggressions at an individual, community, and institutional level.
   - Reporting options, policies, and resources on campus.
   - Acknowledging the role each person has in contributing to the culture of the group which ties into **Rubric Item 2** [civility or respect promotion programs] in promoting civility and respect.
   - Understanding power dynamics and different ways of conceptualizing power which ties into **Rubric Item 3** [leadership education and skill development] by encouraging a broader sense of power beyond positional authority. This is especially important to highlight to graduate students the ways in which they have influence in the community.
   - Bystander intervention skills to recognize and address gender-based harassment and other forms of discrimination which ties into **Rubric Item 4** [bystander intervention programs, specific to higher education or field, and/or audience] by providing participants with different ways of intervening beyond "direct" intervention in the moment.
   - A section on sexual harassment/gender harassment examples and policies which ties into **Rubric Item 5** [audience-specific anti-sexual harassment education] with specific focus on

the difference between ambient harassment and targeted harassment.
- Activities to create more inclusive and welcoming lab environments for everybody, with special attention to the role of faculty members and research staff in setting the tone and holding people accountable, which ties into **Rubric Item 16** [addressing gender harassment and other bad behaviors] by concluding the workshop with an activity that asks labs to come up with the first piece of their values/expectations statement for the lab.

3. **The current status of the work (in progress of finalizing plan/action, currently being implemented, or implemented)**

   These workshops are **currently being implemented**. They have been rolled out successfully in the chemistry department (2018), chemical engineering department (2019), and the Media Lab (spring 2020); we were in the process of working with the mechanical engineering department when COVID-19 disrupted our progress as we went remote on March 16, 2020. We have just tested a virtual version of the workshop with success and are working to develop a way to consistently facilitate the workshops virtually. There are many other departments that are interested in rolling out this workshop, and we are currently working on ways to meet the demand (hiring additional staff, exploring the effectiveness of virtual workshops, etc.).

4. **How this work is either (new or uncommon for higher education)**

   Though the workshop has been continually updated, the actual concept of the workshop was developed pre-2019. The idea of an entire department committing each lab to this workshop is uncommon for higher education, and we think has contributed to its success because it ensures an education dosage for the entire community that is meaningful, interactive, tailored, and occurring during the same time period.

5. **Plans to evaluate the work and/or evaluation results or impact of work**

   We have been regularly evaluating the workshops. Recently, for the Media Lab, we shifted from a more satisfaction-oriented evaluation to a learning outcomes-based evaluation. Additionally, we implemented a 6-month post-test for the chemical engineering department and are in the process of doing so for the Media Lab as well. This post-test evaluation helps us determine the long-term impact of the training.

   Two graduate students in the chemical engineering department helped us submit for publication an overview of the workshop and participant data to the American Society for Engineering Education conference.

6. **How you involved or are involving stakeholders in the plans and/or work**

   We identify three main stakeholders in this initiative: graduate students, faculty of the department, and the chair of the department.

   - Graduate students are often the reason we receive a departmental request, and they are invaluable in helping us customize examples, provide feedback on the flow of the workshop, and share any student-level data they've collected about the graduate student experience in the department.
   - Faculty are important stakeholders because this workshop is done at the lab level and a faculty member's investment and engagement before during and after the workshop impacts the climate and culture of the lab group. One of the ways we specifically engage faculty is by running through the workshop in a faculty meeting with them to incorporate their feedback and create buy-in about the workshop content.
   - Lastly, the department chair is a vital stakeholder in our workshop initiative because they have the ability to communicate to the entire department why it is worth taking time out of our busy schedules to do this and connect it to other work happening at the departmental level to address climate, inclu-

sion, and belonging. As mentioned earlier, the chair also gives faculty an opportunity to vote on the initiative to ensure it is *faculty driven*. We have had some department chairs require it and others strongly encourage it. Participation is higher when it is mandated.

7. **What you envision next steps for this work to be**

Our next steps include working to continually increase our bandwidth to provide workshops for more departments at more regular intervals. One of the options we hope to look into further is a train-the-trainer model of presenting this workshop. Additionally, as time has gone on, labs that took the training in 2018–2019 have asked if there is a second iteration of the workshop. Though we do not have anything formally developed, spending time to think about how new material and content could build on the foundational workshop is another next step.

Lastly, especially in this last academic year, we've received more requests for the content of the workshop to be intersectional in its conceptualization and to not only talk about gender-based harassment but also other forms of micro-aggressions, discrimination, and biased behavior on the basis of race, ethnicity, religion, gender identity, and other categories of identity. Every time we've updated the content we've striven to approach the examples and content in a more intersectional way.

8. **Link to more information about the effort and/or contact info**

Website for further information (if applicable):
Point of Contact Name: Sarah Rankin or Bianca Kaushal
Email Address for Point of Contact: srankin@mit.edu; bkaushal@mit.edu

CASE STUDY F

RUTGERS UNIVERSITY

SIGNIFICANTLY INCREASING FACULTY AND STAFF EDUCATION AND SKILL DEVELOPMENT

This Action Applies to Seven Rubric Items, Five in the Prevention Category and Two in the Response Category: Leadership education and skill development, bystander intervention programs (specific to higher education or field, and/or audience), audience-specific anti-sexual harassment education, ally or ambassador programs, prevention program or toolkits, trauma-informed response and education programs, and addressing gender harassment and other bad behaviors.

Description of Work: We R Here Staff and Faculty Training Initiative

A full-time staff and faculty training coordinator, a position and conceptualization of the work entirely new to Rutgers, was hired at Rutgers in November of 2019 to launch the new We R Here Staff and Faculty Training Initiative across all Rutgers campuses. A core, in-person, anti-sexual harassment training was created. This training, developed with principles of trauma-informed bystander intervention strategies, provides skills to recognize, correct, and address sexual harassment (with a focus on gender-based harassment), support impacted students and colleagues, and effectively use university policies for action and to create positive culture shift.

This interactive training has been tailored for delivery at Rutgers' New Jersey School of Medicine for 700 staff and faculty and will be customized for other university ecosystems accordingly. The We R Here Faculty and Staff Training Initiative will also include the development of a faculty ambassador train-the-trainer program and a comprehensive toolkit with specific, actionable items of change for departments, schools, and academic leaders to adopt to ensure sustainable change.

The goals of the Staff and Faculty Training Initiative are to (1) clearly define sexual and gender-based harassment, (2) discuss how sexual harassment manifests in each specific university environment, (3) provide concrete skills to interrupt sexual harassment in the work place using trauma-informed bystander intervention strategies, and (4) explore concrete action steps to encourage behavior change and to sustainably prevent sexual harassment at Rutgers.

This work aligns with the recommendations of the study *Sexual Harassment of Women: Climate, Culture, and Consequences in Academic Sciences, Engineering, and Medicine* (NASEM, 2018), particularly in the areas of prevention, leadership education and skill development, bystander intervention programs, audience-specific anti-sexual harassment programs, ally and ambassador programs, and prevention toolkits.

As mentioned in the report, the faculty and staff trainings have been designed not to change beliefs but instead to "clearly communicate behavioral expectations" and to provide individuals with the tools to effectively identify, intervene, and prevent sexual harassment both in the workplace and among students. Rather than a one-size-fits-all approach, trainings are specifically tailored to each audience and ecosystem, and are skills-based, interactive, and trauma informed.

Recognizing that training alone cannot bring about lasting culture change, the faculty and staff training coordinator will also create a comprehensive toolkit, which will include best practices for onboarding, sample informal policies and behavioral change measures, trauma-informed resources, sample syllabus statements, and classroom exercises to encourage discussion, social media templates, and departmental and self-assessment tools. A faculty ambassador component is also being developed. In line with the report's findings that women of color are particularly vulnerable to sexual harassment, as well as less likely to report it, each training and intervention has been designed incorporating principles of intersectionality and with an anti-racist, anti-oppressive lens.

This work is currently in progress and continues. Several of the core trainings have been researched, designed, and delivered via WebEx, and outreach to faculty and staff is ongoing. The training coordinator will offer a training series in the fall of 2020 remotely, open to all faculty and staff, that focuses on supporting colleagues and staff remotely during COVID, with a particular focus on Black and other people of color colleagues and students who are disproportionally impacted by the pandemic and systemic racism at large. The series will also feature prominent anti-racist, anti-sexual assault advocate Wagatwe Wanjuki who will focus specifically on supporting Black students remotely.

In addition to offering WebEx and limited in-person trainings for faculty and staff in the coming year, the training coordinator will focus on research, development, and dissemination of the staff and faculty toolkit as well as launching the ambassador program.

It is important to note that certain revisions or changes to the work have taken place, a result of adapting to a remote environment due to COVID-19. The training coordinator worked during March and April (2019) to migrate all trainings to an online platform, although she will still offer limited in-person training to faculty and staff who remain on the ground (for instance, essential medical personnel). Also due to COVID transitions and stressors, demand for training has decreased, but the training coordinator continues to reach out to faculty and staff, including via virtual postcards with action steps and resources, and by offering more training options and making sure content is tailored to shifting needs. A tip sheet for responding to disclosures remotely during COVID-19 was also developed and posted on the university-wide resource site, coronavirus.rutgers.edu, as well as on the university's sexual harassment prevention website, sexualharassment.rutgers.edu.

Assessment is an integral part of the program. Evaluations are provided to each participant after every training, and an online form has been created for WebEx programs. These evaluations will be used to gather feedback and be analyzed for continuous improvement, to ensure that trainings align with Action Collaborative goals [see Chapter 1]. There will also be questions about training and engagement on the upcoming university-wide faculty, student, and staff climate survey, scheduled to be put into the field during the fall of 2021. Interventions and training will be modified accordingly, in response to assessment results.

With regard to involvement of stakeholders in the work, this position itself was developed specifically to engage multiple stakeholders. The training coordinator spends 50% of her time with university human resources, in an effort to streamline training efforts, engage more faculty and staff, and ensure that university policy is appropriately responsive to faculty, staff, and administrator needs around sexual harassment. The training coordinator also works with the leadership of the university's representatives to the Action Collaborative representatives, the Rutgers Center on Violence Against Women, the university Title IX offices, and all violence prevention and victim assistance offices in order to coordinate training, share resources, and remain up to date on university services and policies.

With regard to next steps for the work, in addition to the creation of the toolkit and ambassador program, the next steps will be to continue to respond and adapt to the needs of faculty, staff, and administrators during COVID, including bringing awareness to the fact that sexual and gender-based harassment do not disappear when colleagues and students are working remotely. Since harassment may take different forms and the responses and interventions need to be tailored accordingly, the training coordinator will continue to work to modify training content and offer flexible opportunities as needed. And since all of the aforementioned work is funded by an external grant that ends in August 2021, the training coordinator will continue to explore options for sustainability with university leadership.

Website for further information (if applicable):
Point of Contact Name: Kaylin Padovano, LMSW
Email Address for Point of Contact: kaylin@hr.rutgers.edu

Appendix E

Worksheets for Getting Started with Implementation Science

This appendix offers four approaches for using implementation science in evaluations of sexual harassment.

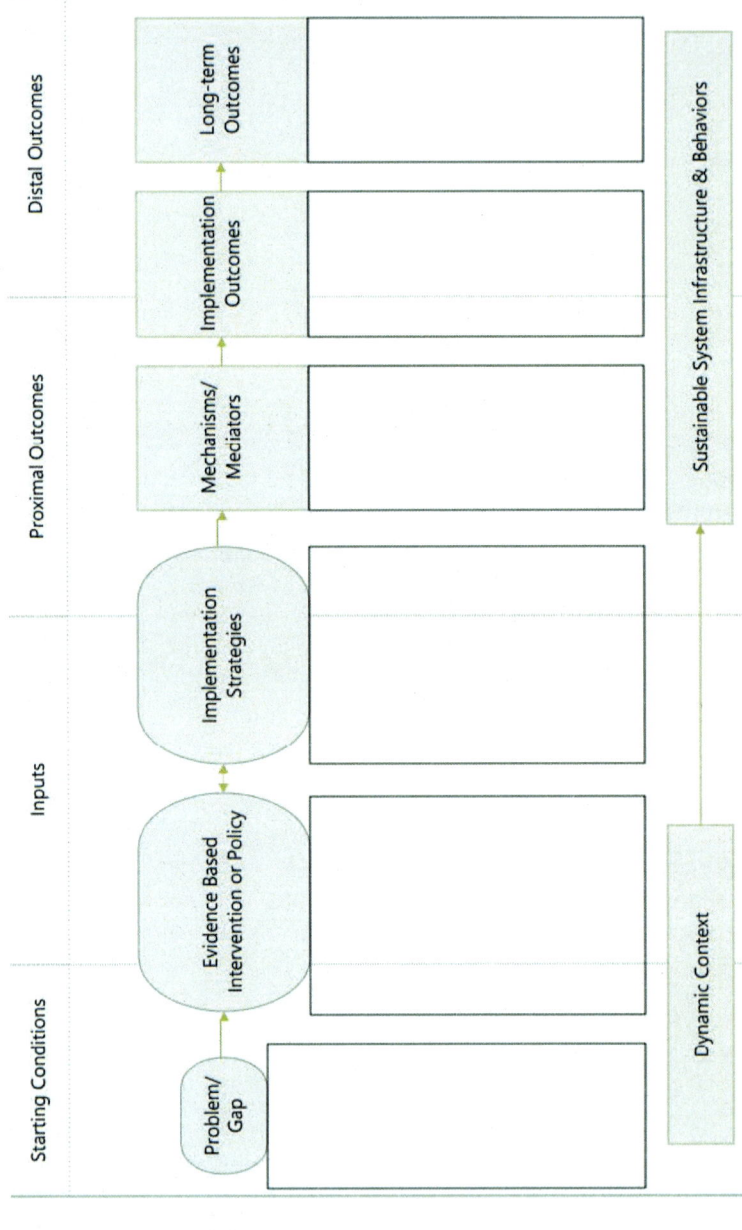

FIGURE E-1 Logic model of an implementation science project.
SOURCE: Adapted from Rabin et al. (2019).

For each item below, select one or more options that describes or is relevant to your project. For each box you check, you should use a separate sheet to add details.

1. **Problem**
   - ❑ Dissemination
   - ❑ Implementation
   - ❑ Policy
   - ❑ Individual Level
   - ❑ Organizational Level
   - ❑ Context – inner setting
   - ❑ Context – outer setting
   - ❑ Target audience

2. **Evidence-Based Program, Intervention, or Policy**
   - ❑ Cost
   - ❑ Relative Advantage
   - ❑ Dose
   - ❑ Acceptability
   - ❑ Trialability
   - ❑ Complexity

3. **Implementation strategies**
   - ❑ Fit
   - ❑ Dose
   - ❑ Compatibility
   - ❑ Champion
   - ❑ Communication channels
   - ❑ Stakeholders

4. **Mechanisms**
   - ❑ Process
   - ❑ Knowledge
   - ❑ Goals
   - ❑ Readiness
   - ❑ Engagement
   - ❑ Transfer

5. **Implementation Outcomes**
   - ❑ Reach
   - ❑ Cost
   - ❑ Adoption
   - ❑ Fidelity
   - ❑ Maintenance
   - ❑ Acceptability
   - ❑ Feasibility
   - ❑ Appropriateness

6. **Long-Term Outcomes**
   - ❑ Intervention outcomes
   - ❑ Fidelity
   - ❑ Maintenance/Sustainability

FIGURE E-2 Logic model worksheet.
SOURCE: Modified from Rabin et al. (2019).

TABLE E-1 Worksheet to Guide Application of the Consolidated Framework for Implementation Research (CFIR)

Not all questions need to be (or even should be) answered. Choose the domains and constructs that make the most sense for your project.

Domain	Constructs	Guiding Questions
Intervention Characteristics		
	Intervention Source	Who developed the intervention? What are stakeholders' opinions of this group/person?
		Why is the intervention being implemented in this setting? Who decided to implement the intervention and how was that decision made?
	Strength of Evidence	What kind of information or evidence are you aware of that shows whether the intervention will work in your setting? Where is that evidence from? How does this evidence influence your perception of the intervention?
		What do administrative or other leaders think of the intervention?
		What kind of supporting evidence is needed to increase stakeholders' buy-in?
	Relative Advantage	What advantages and disadvantages does this intervention have compared to other similar existing programs in your setting?
		How does the intervention compare to other alternatives that may have been considered or that you know about?
		Is there another intervention that people would rather implement? Why do they have this preference?

	Adaptability	What kinds of changes or alterations do you think you will need to make to the intervention so it will work effectively in your setting?
		Are there components that should not be altered?
	Trialability	Can/Will the intervention be piloted prior to full-scale implementation?
	Complexity	How complicated is the intervention in terms of duration, scope, intricacy, and number of steps involved?
	Design Quality & Packaging	What supports, such as online resources, marketing materials, or a toolkit, are available to help you implement and use the intervention?
		How will available materials affect implementation in your setting?
	Cost	What costs will be incurred to implement the intervention?
Outer Setting		
	Stakeholders' Needs & Resources	How "in touch" are staff and leadership with the individuals served by your organization?
		To what extent were the needs and preferences of the individuals served by your organization considered when deciding to implement the intervention?
		How do you think the individuals served by your organization will respond to the intervention?
		What barriers will the individuals served by your organization face to participating in the intervention?
		Have you elicited information from participants regarding their experiences with the intervention?

continued

TABLE E-1 Continued

Cosmopolitanism	To what extent do you network with colleagues or people in similar professions/positions outside your setting?
	What kind of information exchange do you have with others outside your setting, either related to the intervention, or more generally about your profession?
	To what extent does your organization encourage you to network with colleagues outside your own setting?
Peer Pressure	What you know about any other organizations that have implemented the intervention or other similar programs?
	To what extent are other organizations implementing the intervention?
	To what extent are other units within your organization implementing the intervention?
	To what extent would implementing the intervention provide an advantage for your organization compared to other organizations in your area? Is there something about the intervention that would bring more individuals into your organization, instead of another one in your area?
External Policies & Incentives	What kind of local, state, or national performance measures, policies, regulations, or guidelines influenced the decision to implement the intervention? How will the intervention affect your organization's ability to meet these measures, policies, regulations, or guidelines?
	What kind of financial or other incentives influenced the decision to implement the intervention?

Inner Setting		
	Structural Characteristics	How will the infrastructure of your organization (social architecture, age, maturity, size, or physical layout) affect the implementation of the intervention?
		What kinds of infrastructure changes will be needed to accommodate the intervention?
		Changes in scope of practice? Changes in formal policies? Changes in information systems or electronic records systems?
	Networks & Communications	Do you meet (formally or informally) with a team of people?
		Can you describe your working relationship with leaders?
		Can you describe your working relationship with influential stakeholders?
		How do you typically find out about new information, such as new initiatives, accomplishments, issues, new staff, and staff departures?
		When you need to get something done or to solve a problem, who are your "go-to" people?
	Culture	How would you describe the culture of your organization? Of your own setting or unit?
		How do you think your organization's culture (general beliefs, values, assumptions that people embrace) will affect the implementation of the intervention?
		To what extent are new ideas embraced and used to make improvements in your organization?

continued

TABLE E-1 Continued

	Implementation Climate	What is the general level of receptivity in your organization to implementing the intervention?
	Tension for Change	How essential is this intervention to meet the needs of the individuals served by your organization or other organizational goals and objectives?
	Compatibility	How well does the intervention fit with your values and norms and the values and norms within the organization?
		How well does the intervention fit with existing work processes and practices in your setting?
		Can you describe how the intervention will be integrated into current processes? Will the intervention replace or complement a current program or process?
	Relative Priority	To what extent might the implementation take a backseat to other high-priority initiatives going on now?
		How will you juggle competing priorities in your own work? How will your colleagues juggle these priorities?
	Organizational Incentives & Rewards	What kinds of incentives are there to help ensure that the implementation of the intervention is successful?
		To what extent do you think your supervisor will consider your role in this implementation in your (next) evaluation? In their regard for your work or role?
		Are there any special recognitions or rewards planned that are related to implementing the intervention?

	Goals & Feedback	Have you/your unit/your organization set goals related to the implementation of the intervention? (If yes, what are they?)
		To what extent does your organization/unit set goals for current programs/initiatives?
		To what extent are organizational goals monitored for progress?
		How will you get feedback about the implementation?
	Learning Climate	To what extent do you feel like you can try new things to improve your work processes?
	Readiness for Implementation	What level of endorsement or support have you seen or heard from leaders?
		What kind of support or actions can you expect from leaders in your organization to help make implementation successful?
		Do you expect to have sufficient resources to implement and administer the intervention?
		What kind of training is planned for you? For colleagues?
		Who do you ask if you have questions about the intervention or its implementation?
Characteristics of Individuals		
	Knowledge & Beliefs	What do you know about the intervention or its implementation?
		Do you think the intervention will be effective in your setting?

continued

TABLE E-1 Continued

Self-efficacy	How confident are you that you will be able to successfully implement the intervention?	
	How confident are you that you will be able to use the intervention?	
	How confident do you think your colleagues feel about implementing the intervention?	
	How confident do you think your colleagues feel about using the intervention?	
Individual Stage of Change	How prepared are you to use the intervention? • Knowledge stage (Pre-contemplation) - knowledge of key aspects of the intervention • Persuasion stage (Contemplation) - likes the intervention, discusses it with others, buys into it, has a positive view • Decision stage (Preparation) - intends to seek additional information and try it • Implementation stage (Action) - acquires additional information, uses intervention regularly, and has continued use • Confirmation stage (Maintenance) - recognizes benefits, has integrated the intervention into routines, promotes use to others	
Individual Identification with the Organization	What are individuals' buy-in to organizational or intervention-related goals?	
Other personal attributes	Other motivational or behavior change constructs that could be related to the intervention.	

Process		
	Planning	What have you done (or what do you plan to do) to get a plan in place to implement the intervention?
		Can you describe the plan for implementing the intervention?
	Engaging	Who are the key influential individuals to get on board with this implementation?
		What are influential individuals saying about the intervention?
		How did your organization become involved in implementing the intervention?
		Other than the formal implementation leader, are there people in your organization who are likely to champion (go above and beyond what might be expected) the intervention? Can you describe people's perception of this champion/individual?
		Will someone (or a team) outside your organization be helping you with implementing the intervention?
		What steps have been taken to encourage individuals to commit to using the intervention?
		Who are the key individuals to get on board with the intervention?
	Executing	(During or post-implementation)
		Has the intervention been implemented according to the implementation plan?

continued

TABLE E-1 Continued

Reflecting & Evaluating	What kind of information do you plan to collect as you implement the intervention? Which measures will you track? How will you track them? How will this information be used?
	Will you receive feedback reports about the implementation or the intervention itself?
	How will you assess progress toward implementation or intervention goals? How will results of the evaluation be distributed to stakeholders?

SOURCE: Modified from the Consolidated Framework for Implementation Research. Available: https://cfirguide.org/guide/app/#/.

TABLE E-2 Worksheet to Guide Application of the Practical, Robust Implementation and Sustainability Model (PRISM)

Not all questions need to be (or even should be) answered. Choose the domains and constructs that make the most sense for your project.

Domain	Constructs	Guiding Questions
Program/Intervention		
	Organizational Perspective	Are there any specific staff roles that are needed to support the intervention? Are these already in place or do they need to be created?
	Consumer Perspective	How does the intervention affect the intended target(s)?
		How do the target(s) of the intervention feel about the current situation (pre-intervention)?
External Environment		To what extent does the intended intervention fit in with national priorities in higher education?
Implementation and Sustainability Infrastructure		What does the ideal setting (e.g., lab, department, college) look/function like? What resources need to be in place to make this happen?
		How would the intervention continue to persist in the workflow of the organization?
		Do you have the tools and skills needed to continue to provide and develop the intervention?

continued

TABLE E-2 Continued

Recipients	
Organizational	Is sexual harassment prevention explicitly valued in the organization? For the individuals within the organization?
	How much support is there from institutional leaders to combat this issue? Are there any champions to support the intervention?
	Does the intervention fit in with institutional priorities?
	What processes or policies are already in place around this issue? How could they affect implementation of the intervention?
	What barriers currently exist for addressing sexual harassment in your organization?
	What are the benefits to the organization of participating in implementation?
Consumer	How confident are you that you will be able to successfully implement the intervention?
	How confident are you that you will be able to use the intervention?
	How confident do you think your colleagues feel about implementing the intervention?
	How confident do you think your colleagues feel about using the intervention?
	What is your opinion of how the institution handles sexual harassment and prevention? What are your concerns about these topics? How are you affected by current policies? How would receiving the intervention impact you?

SOURCE: Modified from McCreight et al. (2019).